Complexity and Public Policy

Complexity and Public Policy: A New Approach to 21st Century Politics, Policy and Society challenges the predominantly orderly paradigm, concepts and tools of public policy with a new framework based on a complexity perspective, concepts and tools.

This book provides a clear, concise and readable introduction to complexity thinking, its application to the social sciences and public policy, and the relevance of some of its various tools to those fields of politics, health, the international realm, development, planning and terrorism. The authors argue that the foundation for many of the current crises in these areas can be traced to the attempt by social scientists and policy-makers to treat these systems and processes as fundamentally orderly, predictable and controllable. Adopting a complexity framework and using some its tools is a first step in moving towards a more realistic and humane politics, policy and society in the 21st century. The book is vital reading for students and scholars of the social sciences and public policy, and also policymakers and the policy actor audience.

Robert Geyer is Professor of Politics, Complexity and Policy at the Department of Politics and International Relations at Lancaster University, UK.

Dr Samir Rihani is Research Fellow at the University of Liverpool, UK and is Non-Executive Director of the Liverpool Primary Care Trust, UK.

Complexity and Public Policy
A new approach to twenty-first century politics, policy and society

Robert Geyer and Samir Rihani

LONDON AND NEW YORK

First published 2010
by Routledge
2 Park Square Milton Park Abingdon Oxon OX14 4RN

Simultaneously published in the USA and Canada
by Routledge
270 Madison Avenue, New York, NY 10016

Routledge is an imprint of the Taylor & Francis Group,
an informa business

© 2010 Robert Geyer and Samir Rihani

Typeset in Times New Roman by Glyph International Ltd.
Printed and bound in Great Britain by TJ International Ltd, Padstow

All rights reserved. No part of this book may be reprinted or reproduced or utilised in any form or by any electronic, mechanical, or other means, now known or hereafter invented, including photocopying and recording, or in any information storage or retrieval system, without permission in writing from the publishers.

British Library Cataloguing in Publication Data
A catalogue record for this book is available from the British Library

Library of Congress Cataloging-in-Publication Data
Geyer, Robert.
Complexity and public policy : a new approach to 21st century politics, policy, and society/Robert Geyer and Samir Rihani.
    p. cm.
Includes bibliographical references.
1. Policy sciences. 2. Political planning. 3. System theory. 4. Complexity (Philosophy) 5. European Union countries–Social policy–Planning.
6. Terrorism–Prevention. I. Rihani, Samir, 1938- II. Title.
H97.G49 2010
320.6–dc22   2009034892

ISBN 10: 0-415-55662-7 (hbk)
ISBN 10: 0-415-55663-5 (pbk)
ISBN 10: 0-203-85692-9 (ebk)

ISBN 13: 978-0-415-55662-0 (hbk)
ISBN 13: 978-0-415-55663-7 (pbk)
ISBN 13: 978-0-203-85692-5 (ebk)

Contents

List of figures ix
List of tables xi
Preface xii
Acknowledgements xiv

Introduction 1
*The problem of order: three symbolic cases – Nicolas,
   Emile and Leila 1
From the pursuit of order to the paradigm of complexity 5
Three brief points and a note on definitions 7
The plan of the book 8*

1 **From orderly to complexity science** 12
*The paradigm of order 12
Spreading ripples of doubt 14
Complex systems in the physical world 16
Complex systems in the biotic world 18
Orderly (modernist) social science and public policy 20
Disorderly (post-modernist) social science
   and public policy 24
Complexity and social science 28
Complexity and public policy 30*

2 **Concepts of complexity** 36
*Key concepts of physical complex systems 36
Key concepts of biotic complex systems 41
Key concepts of conscious complex systems 46*

3 **Tools of complexity** 53
Cascade of complexity 53

Balance and range of outcomes 55
Complexity mapping 57
Fitness landscape 60
A different point of view: the fitness landscape 61
The Stacey diagram 64
Stakeholder involvement and soft systems methodology 68

4 Politics 73
What would a 'complexity map' of politics look like? 74
Complexity and democracy 76
What would a complexity theory of democracy look like? 77
Fundamental regularities and threats 78
Complexity and the 'third way' 80
How does complexity differ from and go beyond the third way? 84
Complexity, politics and the third sector 85
Is the EU compatible with the third sector? 88
Using the 'balance and range of outcomes' tool to help the EU promote the third sector 89
Conclusion 90

5 Health 92
The influence of the paradigm of order on health 92
Better health at a fraction of the cost – focusing on the basics 94
An efficient national health system: the Cuban case 95
Is Cuba's experience unique and what is the magic formula? 96
The case of the English national health service 98
Complexity and chronic illness: the case of diabetes 103
Understanding chronic illness – from order to complexity 104
Using a fitness landscape to improve diabetes management 105
A different point of view: the fitness landscape 106
Bringing it all together 109

6 The international arena 110
International relations theory 111
European integration theory 112
Integrating complexity 113
Applying a complexity map to the international arena and European Union 114
General implications 116
Globalisation, Europeanisation and the curious case of the non-death of Scandinavian exceptionalism 118

Implications 120
*Another example: the predictable proposal for a true
  capital of Europe 122*
*Using a complexity cascade to visualise the development
  of the EU, Scandinavia and Brussels 123*
Conclusion 128

7 **Development** 129
  The case for radical change in development practice 130
  Lopsided view of the way nations evolve 132
  Why to the successful succeed? Self-organised complexity 133
  Why do the unsuccessful stagnate? 134
  A more realistic vision of development 136
  From x–y to fitness landscape thinking 138
  Why changing the framework of development is so slow 142
  Concluding remarks: from hard to soft management 143

8 **Planning dreams into nightmares: The Iraq adventure** 146
  The predictability of failure 147
  Doubts were ignored 149
  Was it just state incompetence and/or lack or preparation? 151
  Massive costs and negligible benefits 152
  The underlying questions 154
  Where did the complexity come from? 156
  *Three acts of orderly reasonableness, but complexity
    madness 158*
  Could others have done better? The Stacey diagram 160

9 **Exploding the myths of terrorism** 164
  *Myth 1: There is a clear and agreed definition of terror, terrorism
    and terrorist 165*
  *Myth 2: September 11 was a tragedy and not a political
    opportunity 166*
  *Myth 3: One can reduce the elements of a complex international
    system into separate and manageable units 168*
  Myth 4: The 'war on terror' is a global affair 169
  *Myth 5: Global and regional powers are simply reacting to
    terrorism 169*
  Myth 6: The 'war on terror' is being won 170
  *Myth 7: Current terrorism is a new type of
    terrorism 172*

Myth 8: The scale of terrorism has radically increased 173
Myth 9: Middle eastern and Islamic terrorism is a new and major threat 173
Myth 10: Al-Qaeda is new as well 177
Beyond the myths: what does complexity tell us? 178
A complexity map of international terrorism and its implications 179

10 Conclusion: Towards a complex and humane public policy for the twenty-first century   183
Isn't complexity just a recipe for doing nothing? 183
Is there a morality of complexity? 184
What can complexity do for the weak? 185
Why has complexity been so slow to spill over into the social sciences and public policy? 186
Can complexity be inspirational? 187
How do we go forward from here? 187
And finally ... a couple of playful mental exercises 188

Notes	192
Bibliography	199
Index	211

Figures

1.1	Phenomena in the paradigm of order	14
1.2	Phenomena in the paradigm disorder and order	15
1.3	The range of physical phenomena in a complexity paradigm	18
1.4	The range of physical and biotic phenomena	19
1.5	The range of physical, biotic and conscious phenomena	29
3.1	Complexity phenomena and the arrow of time	54
3.2	Range of outcomes	56
3.3	Complexity mapping for human phenomena	57
3.4	The range of complexity in everyday life	59
3.5	Complexity map of President Kennedy's decision to use a naval blockade to stop further Soviet missile development in Cuba	60
3.6	Typical x–y graph	61
3.7	Example of a fitness landscape	62
3.8	Fitness landscape combined with balance and range of outcomes	63
3.9	Stacey diagram step 1	65
3.10	Stacey diagram step 2	67
3.11	Traditional decision-making process. Based on an original image at http://www.metronetiq.com/archives/2007/09	70
3.12	Traditional decision-making overlaid with real decision-making process based on an original image at http://www.metronetiq.com/archives/2007/09	70
3.13	SSM inspired organisational strategy plan. Used in the NHS improvement Leaders' Guide, *Working in Systems (2005)* but originally published in langley G, Nolan K, Norman C, Provost L, (1996). The Improvement Guide: a practical approach to enhancing organisational performance, Jossey Bass Publishers, San Francisco	71
4.1	The range of complexity in politics	74
5.1	An orderly perspective: the x–y graph and blood glucose control	106
5.2	Visual representation of a fitness landscape of strategics for managing diabetes. Drawing created by Laura Fleming, http//laurafleming.co.uk/html/home.htm	108

x *Figures*

6.1	The range of international arena phenomena	114
6.2	A complexity map of the European Union	115
6.3	The traditional model of unity and order	124
6.4	European Union complexity cascade	125
6.5	Scandinavian social democratic complexity cascade	126
6.6	Brussels as a capital for Europe complexity cascade	127
7.1	United Nations Development Programme, *Human Development Report*, (1999: 38)	131
7.2	An x–y view of development	138
7.3	A fitness landscape view of development	139
7.4	Another fitness landscape view of development	140
8.1	A Stacey diagram of the restructuring of Iraq	161
9.1	Total terrorist attacks, 1981–2001. (This chart is copied from Appendix 1, Statistical Review, Pages 171–176 of the *2001 Patterns of global Terrorism Report*)	174
9.2	Total international casualties by region, 1996–2001. (This chart is copied from Appendix 1, Statistical Review, pages 171–176 of the *2001 Patterns of Global Terrorism Report*)	175
9.3	Total international attacks by region, 1996–2001. (This chart is copied from Appendix 1, Statistical Review, pages 171–176 of the *2001 Patterns of Global Terrorism Report*)	176
9.4	A complexity map of international terrorism	179

Tables

1.1	The foundations of orderly (modernist) social science	22
1.2	The foundations of orderly (modernist) public policy	24
1.3	The foundations of disorderly (post-modern) social science	26
1.4	The foundations of disorderly (post-modern) public policy	28
1.5	The foundations of a complexity social science	31
1.6	The foundations of a complexity public policy	32
1.7	Summaries of orderly, complex and disorderly public policy perspectives	33

Preface

Who are we and why did we write this book?

This book emerged out of two related frustrations in the 1980s: Robert Geyer's inability to find 'true' socialism and Samir Rihani's despair at the plight of poorer nations to get off the bottom of the ladder in a harsh and baffling world.

For Robert, an idealistic young American who was disgusted with the limitations and right-wing orientation of American politics during the Reagan presidency, Europe offered multi-party democracy, real socialist parties and a chance to check out the edge of the Soviet bloc. After studying in Paris and England, he returned to the US and began his PhD at the University of Wisconsin determined to solve the crisis of democratic socialism. However, as the PhD progressed he became less and less convinced of the intellectual hegemony of socialism. It still held his heart and hopes, but couldn't explain the amazing transformations at the end of the twentieth century. Moreover, as his interests took him deeper and deeper into the mire of European Union politics, it became increasingly clear that no theory could possibly deal with its evolving multi-level dynamics in a traditional scientific fashion. In essence, he had finally realised the remarkable complexity of social and economic life and began to ask, 'What next?!'.

Samir grew up as a Christian in Baghdad and left to study for a civil engineering degree at the University of Liverpool. After obtaining his degree, he married and settled in Liverpool working, among other places, in the planning and transport departments of Liverpool's local government. Samir was part of Liverpool's largest planning exercise when Liverpool and its sub-region tried to 'plan' its way out of its economic and demographic difficulties in the 1970s. The ink was barely dry on the plan when it became obvious to Samir that this massive exercise was a complete waste of time and effort, a desperate attempt to put a gloss of order on a fundamentally chaotic situation. He also reached a broader conclusion: if they can't get it right in Britain what chance is there for poorer counties to move ahead? Later, after working in tourism and international development Samir went back to the University of Liverpool to undertake research for a PhD in complexity and international development. Robert, a young academic at the time, was one of Samir's PhD supervisors.

In the beginning, Robert was very sceptical of Samir's thinking on complexity. Without a physical science background, Robert was afraid to make 'unsafe' linkages to those non-social science fields. Moreover, as a young academic at the beginning of his career he was scared of going beyond the safety of his particular area of expertise. His own training and academic experiences had shown him that the best way to get along in academic circles was to pick a small research topic and focus on it exclusively. Over time he would become the expert on that area, a safe pair of hands.

But Samir would not let up. Week after week he would confront Robert with new complexity ideas and concepts. As the weeks turned into months, then years, Robert's reluctance and fear began to fade. The complexity literature in the physical and social sciences continued to build up and Robert began to apply it to his areas of expertise, European Union politics and policy, Scandinavia and public policy. It gave him a new way of thinking about old problems and a broader perspective that refused to allow him to go back into his safe box. And most wonderful of all, it made academic work fun again! The obvious next step was some sort of joint project, or book.

The first signs did not auger well. Samir was busy writing his book, *Complex Systems Theory and Development Practice*, lecturing at the School of Politics at the University of Liverpool, building up his website www.globalcomplexity.org and acting as a non-executive director on various National Health Service Boards in Liverpool. Robert was also busy with normal academic demands, publishing several books, creating a complexity network at Liverpool and running a major 'Complexity, Science and Society' conference in 2005. In 2006 he moved to Lancaster University. Nevertheless, the book was always in the back of their minds. They just needed to get enough time to bring the multiple strands together. What it finally took was one swift kick up the backside by the best student they had ever taught, Abbie Badcock. The result is before you.

Something that has taken us so long to put together obviously generates a long list of people to thank. For work colleagues, in the Department of Politics and IR at Lancaster University we would like to thank Patrick Bishop, Patricia Chilton, Mick Dillon and Lee Miles for commenting on drafts and ideas. Others at Lancaster who inspired our work include: Bob Jessop, John Urry, Sylvia Walby and Graham Harris. For other colleagues around the UK and world we would like to thank Peter Allen, Jan Bogg, Cary Brown, Ceri Brown, David Byrne, Fritjof Capra, Paul Cilliers, Barry Cooper, Helen Cooper, Andrew Deakin, John Dearing, John Lewis Gaddis, Carlos Gershenson, Tim Holt, Rod Lambert, David Marsh, Peter McBurney, Will Medd, Eve Mittleton-Kelly, Mateo Willis and Graham Wilson for ideas and inspiration. For current and former students we would like to thank Abbie Badcock (of course), Chris Lawson, Kai Lehmann, Steve Pickering, Jim Price and Stephen Royle. We would also like to thank our families for their continual love and support throughout the 'emergence' of the book.

Acknowledgements

Finally, due to the rather lengthy and bumpy path that this book has had to follow from idea to print several sections are based on earlier publications. They are all updated, edited and amended, but we need to recognise their roots and thank the publishers (Radcliffe Publishing, Blackwell Press, and Europeanvoice) for allowing us to use these earlier works. Several sections of Chapter 1 are based on Geyer, R. and Mackintosh, A. (2005) *Integrating UK and European Social Policy: the Complexity of Europeanisation*, Oxford: Radcliffe Publishing. Three sections in Chapter 4 are rooted in Geyer, R. (2003) 'Beyond the third way: the science of complexity and the politics of choice' *British Journal of Politics and International Relations*, 5 (2): 237–57. Four sections in Chapter 5 are founded on Cooper, H. and Geyer, R. (2007) *Riding the Diabetes Rollercoaster*, Oxford: Radcliffe Publishing. Chapter 6 is derived from elements of Geyer, R. (2001) 'Europe needs a capital city with difference' *Europeanvoice*, 7 (33): 13–19 September; Geyer, R. (2003) 'European integration, the problem of complexity and the revision of theory', *Journal of Common Market Studies*, 41 (1): 15–35; and Geyer, R. (2003) 'The end of globalisation and Europeanisation, rise of complexity and future of Scandinavian exceptionalism', *Governance*, 16 (4): 559–77. Lastly, some of the ideas were first explored in unpublished articles in www.globalcomplexity.org.

Introduction

The problem of order: three symbolic cases – Nicolas, Emile and Leila

Nicolas was a farmer from a prosperous agricultural region, south of Moscow. His family had lived on the farm for generations and were tightly linked to the life of their village. Successive attempts by Russian governments at the end of the nineteenth and early twentieth centuries to modernise the Russian peasantry had only marginally affected Nicolas and his family. He was a soldier during WWI and passively supported the Communists, not because he believed in the cause but because they were the only ones who promised to stop the war. After the 1917 revolution he returned to his family and farm and managed to prosper despite the social and economic turbulence that surrounded him. Then collectivisation began.

Nicolas had originally been happy when Comrade Stalin had become leader of the Politburo. He seemed to promise security and stability. Backed by a clear vision of the future and a raft of experts, Stalin was determined to solve the problems of Russian agriculture through mass collectivisation. It was a rational end-state plan. Russian agriculture had always been undercapitalised, small-scale and decentralised. In one radical stroke the state elites would concentrate all agricultural capital, extend centralised state control and increase efficiency, enabling them to pay for their huge industrialisation drive of the 1930s.

Like his neighbours, Nicolas did his best to survive. He hid his tools, slaughtered and ate his animals and tried to be loyal. However, because he had refused to allow his daughter to marry the local butcher, a man Nicolas had never liked, the butcher denounced him as a *kulak*, a wealthy peasant who opposed collectivisation and the Communist government. Nicolas was tried by the local party functionary and sentenced to spend the remainder of the 1930s in a Siberian work camp along with millions of others. Before his exile he was able to steal a handful of soil from his farm and promise himself that he would return. When he died five years later of cold and exposure, his pocket still held a few grains of his precious soil.

Emile thought she was protected by a lucky wind. Born on the election day of the first black Zambian government in 1962 her youth seemed full of potential. Through the creation of new government schools and careful saving by her parents, she became the first member of the family who had ever learned to read and write. Her teachers, particularly old Miss Mbkei, saw that she had intelligence and encouraged her to continue. At the age of 19, newly married to Kenneth, a small shopkeeper, she got a job teaching English at a new school on the outskirts of the Zambian capital, Lusaka. She thought of having children but was determined to wait until she and Kenneth could afford their own home with a radio and maybe even indoor plumbing. They both worked very hard, Emile gave private lessons in the evening in their room over the shop. They started to save.

However, a new kind of wind began to blow. With the collapse of the price of copper during the 1970s, the Zambian government's export earnings began to evaporate. All of the government experts agreed the downturn would be short and since international interest rates were low they would just borrow a little money to get over the difficulty. During the late 1970s and early 1980s, the small debt turned into a mountainous burden and the International Monetary Fund/World Bank stepped in to 'structurally adjust' the country. All of the Western experts agreed, if Zambia could develop free markets, it would flourish. Emile and Kenneth felt the new markets immediately. Their food costs trebled. Emile could no longer find private students to tutor and her pay slips began to come later and later as state funds vanished. During the first food riots in the mid-1980s the shop was looted.

Kenneth was stabbed to death defending the shop during the second wave of riots. A year later, Emilie's pay checks stopped coming. She started charging the few students she had left. As these numbers declined, she couldn't afford her room and was forced to move to a shanty town. With no family and no connections in the shanty, she was easy prey for the organised gangs. Threatened with starvation, she fell into prostitution. She got pregnant. The child died a year later of diarrhoea brought on by drinking polluted water. Like millions of others she knew the water was bad, but had no way to clean it or buy clean water. She became pregnant again and died in childbirth. A dry wind blew over the rubbish tip where her body was left to rot.

Leila never liked politics. Growing up in post-colonial Iraq, she did her best not to pay attention to the political transformations that were going on around her. Though most Iraqis were frustrated by the slow rate of development and dismayed by successive governments intent on imposing 'order, freedom and democracy' in a bid to accelerate the country's progress, to Leila her family was her world. She used to enjoy laughing in the faces of political groups who would ask for money to support their cause. She married well, had three healthy children and led a privileged life in urban Baghdad, spending the hot summers in their vacation home in the beautiful and cool mountains in the north. Her children were well educated in Iraq and were sent abroad to attend university. To Leila's irritation, all three of the children decided to stay. Leila did her best to travel to see them (and her grandchildren) especially during the hot summer months.

The long years of the Iran–Iraq war were a tragedy for millions, but only an inconvenience for Leila. Her husband's international trading business continued to do well and her children and grandchildren were safely out of the country.

She could even travel when she wanted. When the war ended, life briefly returned to normal. And though her husband died in 1990, he left her with a large amount of savings.

The serious problems began with the UN sanctions. Saddam Hussein had stupidly invaded Kuwait. After the fighting, the UN had imposed the sanctions on Iraq in order to punish Saddam. Within six months, the hundreds of thousands of Iraqi Dinars in her bank account, which had been worth hundreds of thousands of US dollars, were now worth a few hundred dollars. Still, she knew she was lucky. As more and more around her suffered, particularly children, she was able to survive thanks to the money her children were sending from abroad. With the money, she could buy what she needed on the thriving black-market. What galled her was that because of his control of the smuggling routes 25–50% of the cost of black-market goods was going straight to Saddam Hussein!

When the US invaded in 2003, Leila was confident that life would quickly get back to normal. At first it all went well. Saddam's regime collapsed. The Americans took over and she even hoped to travel out of the country to see her grandchildren. But then, despite the declared intentions of the US and its allies, the chaos came. Soon, she was afraid to leave her house at night, then during the day. The money sent by her children was being intercepted. She began to starve. In a desperate attempt to get out, she fled to Syria but never made it to the border. The smuggling gangs dropped her body in the desert.

These are not real stories, but imaginary cases that represent the lives of millions of people. Nicolas symbolises between 10–20 million people who died as a result of the Soviet agricultural collectivisation drive of the 1930s. In Emile's case, the average life expectancy of a Zambian fell during the period of IMF/World Bank 'structural adjustment' from 53 to 41 years (United Nations Develpoment Programme 2002).[1] For a society of 12 million, this represents the loss of 144 million years of life, a high price to pay for the glorious goal of a 'free market economy'. Finally, for Leila, the UN estimated that over a half a million children in Iraq died as a result of the UN sanctions. At the same time, *Fortune* magazine calculated that Saddam Hussein, before the US invasion in 2003, had become the seventh wealthiest man in the world primarily through his control of smuggling routes into Iraq. Meanwhile, during the chaos in Iraq that followed the US invasion, the esteemed British medical journal, *The Lancet*, estimated that by June 2006 over 650,000 had died in the conflict while the UN reported that four million had fled Iraq or were internally displaced.

What links these unfortunate events? Were the people caught within them just unlucky, at the wrong place at the wrong time? Or was something more commonplace and fundamental at work? Despite their being caught in different systems, times and places, they all represent the plight of being a complex person in a society in which brutal, and often well-intentioned, internal or external decision-makers were attempting to impose the impossible: a rigid orderly outcome on an inherently complex adaptive situation.

If the history of public policy in the twentieth century is anything, it is the pursuit of the perfection of greater order on messy societies. Greater order brought huge benefits during the industrial age (economic growth, mass education and health provision, etc.). However, the belief that more order was always better and that some sort of 'final (new) order' could be created and locked into place led to the renowned historian Eric Hobsbawm labelling the twentieth century as the 'Age of Extremes' (Hobsbawm 1994). The extremes were caused by the pursuit of and competition between different visions of human order. Like George Orwell's nightmare vision of the future in 1984, the heartless methodical vision of rational order came cracking down again and again on the human face of complex twentieth century society. The shape of the boot changed, but the stomping remained the same. For the fortunate few who fit the mould, the system – whether communist, fascist or free market – delivered reasonable if not exceptional rewards. For the rest, suffering, degradation and death were *common* outcomes. Often, those in power and who benefited from the system saw the costs of imposing their visions of order upon others as wholly acceptable. As the repetitive rationalisation went throughout the twentieth century, 'the ends justify the means' and the 'new order' (whether announced by the Ottoman Empire, Nazi Germany, Imperial Japan, the Soviet Union, or post-modern America) will bring a new dawn.

The pursuit of order didn't always take such extreme forms nor has it declined in the twenty-first century. As later chapters will demonstrate, a belief in the creation of a final stable order distorts the policies and lives of millions in wealthier countries as well. From legal systems that demand uniform punishments to complex and uncertain crimes, to health targets that force doctors to treat patients like machines, to teaching philosophies which assume that all children develop uniformly, orderly solutions to complex problems infect our politics and daily lives. This is a messy contradiction for the social scientist, but an extremely stressful position for the public policy actor – caught between the demands of orderly, rational central criteria and the messy reality of day-to-day local conditions and contradictions. Moreover, recent twenty-first-century events have certainly confirmed the surprises and pitfalls foisted by a complex world on decision-makers locked into an orderly frame of mind continue to make their presence felt. Apart from the disastrous aftermath of the Iraq war and the increasingly discredited 'war on terror' (two events that materialised seemingly out of nowhere to confound one and all), the deep recession in 2008–9 that wreaked havoc with the elaborate world economic setup and the more 'local' issue of the current MP expenses scandal in the UK that erupted in spring 2009 demonstrate how social scientists and policy actors can not afford to continue to blindly ignore uncertainty, unpredictability and underlying complexity.

One of the central messages of this book is that from a complexity perspective, the variegated pragmatism of less 'orderly' approaches and policy actors, often denigrated as being 'soft', 'non-scientific' and 'woolly' are actually just as scientific as traditional orderly approaches and provide the flexibility, adaptability and sustainability to keep societies from drifting towards the more extreme perils of twentieth-century order. Complexity provides the scientific foundation for

Introduction 5

understanding these 'common sense' implications and some of the tools for avoiding the pitfalls of too much order and disorder.

From the pursuit of order to the paradigm of complexity

The pursuit of order, a desire to comprehend and predict our own world and condition, has been a major part of all recorded human history. For millennia, religion fulfilled the role of providing order in an unknowable universe. The dominant belief in the knowable order of the world and rational and orderly nature of human existence only emerged in the Western philosophical tradition in the Renaissance and Enlightenment. Present critics of human rationality such as post-modernists, constructivists, and so on, are easily found and generally accessible. Nevertheless, a belief in the orderly nature and fundamental rationality of society and the ability of traditional scientific endeavour to understand and direct society remains the bedrock of modern natural and social sciences and public policy.

As we will demonstrate more thoroughly in Chapter 1, based on a Newtonian[2] vision of an orderly, clockwork universe driven by observable and immutable laws, it did not take much of an intellectual leap to apply the lessons of the physical sciences to the social realm. Adam Smith and David Ricardo claimed to have captured the laws of economic interaction. Karl Marx wedded his vision of class struggle to an analysis of the capitalist mode of production to create the 'immutable' and deterministic laws of capitalist development. Political philosophers such as John Stuart Mill and other utilitarians atomised the world by viewing individuals as rational self-interested actors, while August Comte and the positivists created a vision of scientific society based on order, laws and progress.

With the radical expansion of state powers and exponential growth of international interaction in the twentieth century, the pursuit of human order deepened, expanded and led to the extreme forms of suffering discussed above. The repression, death and suffering which the Soviet peoples experienced, particularly during the 1930s, in the pursuit of a Communist order are mirrored in the repression, death and suffering brought on the Third World by the World Bank/IMF's pursuit of a free market order. In other words, the twentieth century was not only the 'Age of Extremes' as the title of Eric Hobsbawm's book suggested, but the 'Age of Unattainable Orders'.

In the 1950s, 60s and 70s, this traditional positivist approach permeated the social sciences and public policy thinking to a particularly high level from the modernisation theories of Third World development, the realist vision of international relations, the functionalist interpretation of European integration, to the rational end-state plans of urban planners. The 1980s and 1990s emphasis on 'centralised public management' and the 'targeting and audit culture' are clear developments of these earlier trends. Nevertheless, despite the collapse of the 'laws' in all of these theories and the continued negative impacts of these rigid orderly approaches, social scientists and policy actors are still caught between the desire to seek out new 'laws' or the partial abandoning of the current framework. This debate has laid the groundwork for a Kuhnian 'paradigm shift'[3] as social scientists and policy actors, increasingly dissatisfied with this pursuit of order, ask where do we go from here?

A theory for moving beyond the limitations of the current Newtonian frame of reference has, ironically, emerged from the physical sciences and is generally labelled 'complexity theory' or just 'complexity'.[4] During the latter half of the twentieth century, scientists continued to find physical phenomena that were not amenable to the traditional Newtonian scientific method. Examples included weather patterns, natural evolution, neural networks, the behaviour of metals under extreme temperature conditions, and quantum uncertainty.

As we will explore more thoroughly in Chapter 1, at this time the physical sciences began to distinguish between 'linear' and 'non-linear' phenomena. From a linear or orderly perspective, causes lead to known effects in a predictable and repeatable manner. Systems could be disassembled to understand the behaviour of their constituent elements and then reassembled, clockwork fashion, to model the behaviour of the whole system under differing conditions. From a non-linear or complexity viewpoint, systems are composed of numerous elements that interact locally according to simple rules to maintain simultaneously massive internal variety and global stability. The internal dynamics of the system create complex outcomes that are not amenable to precise prediction. In general, these phenomena clearly reflect the uncertainty and complexity of the majority of social phenomena and experiences. Reflecting its growing influence in the physical sciences, complexity theory has begun to spill onto the edges of the social sciences, particularly in the field of economics and business management.[5]

We contend that due to the continued pervasiveness of the orderly paradigm and the Newtonian foundations of the institutional structures of the social sciences and public policy, modern social scientists and policy actors unjustifiably continue to assume that social phenomena are primarily orderly and therefore controllable and predictable to a high degree of detail. They, consequently, apply reductionist methods based on the belief that stable relationships exist between causes and effects, such as the assumption that individual self-interest is an explanation and/or a model for national level self-interest. This inability to modify the Newtonian frame of reference explains both the continued failure of social scientists to capture the 'laws' of social interaction and policy actors' continual frustration over their inability to fully control and direct society.

In essence, we believe that the natural-science-based Newtonian paradigm significantly shaped nineteenth- and twentieth-century social sciences, public policy and the societal pursuit of human order. Since nature was orderly, linear and rational, society should be as well. It was just a matter of time, effort and will before the true social laws were found and the end-state society was engineered to bring humankind's history to an end. The horrors of order were the obvious result. Similarly, we hope that the complexity paradigm that emerged in the natural sciences in the twentieth century will greatly influence the shape of the social sciences in the twenty-first century. We want to encourage this transformation because we believe that it lies at the foundation of making the twenty-first century more humane and peaceful than the twentieth.

Three brief points and a note on definitions

As we have presented our ideas over the past few years we have been constantly confronted by three main criticisms: the orderist's grumble, the disorderist's whinge and the 'Where's the beef?' question. The first two are academic criticisms made by the defenders of the Newtonian paradigm who complain that we fail to recognise the importance of the orderly paradigm's achievements and its continued potential on the one hand (the orderist's grumble), and the opponents of the Newtonian paradigm who protest that we do not go far enough in our attacks on the orderly vision of the world (the disorderist's whinge). This book is not an anarchist's cookbook, nor is it a defence of the orderly nature of the natural and social worlds. As we will argue, if one accepts the conclusions of complexity, one must accept that the natural and social worlds are symbiotically intertwined and that they exhibit orderly, disorderly and complex phenomena. On this fundamental fact the entire structure of complexity stands.

The "Where's the beef?" question is the common complaint of the non-academic. In other words, how does complexity thinking affect my particular policy area? Complexity does not have a specific final 'answer' to a particular policy, economic or social issue, or the general woes of the world for that matter. Adopting a complexity framework enables decision-makers to interpret what goes on in the social, economic and political arenas in a new way that recognises the limits of knowledge and prediction and the consequent need to adjust policy-making and actions accordingly. For example, soft management methods, as described later, replace the outwardly forceful but practically blunt traditional hierarchical hard management methods. Ephemeral certainty is abandoned in favour of honest recognition of sensible limitations. The applied chapters of the book are all about trying to spell out the implications of this realistic vision on political, social and economic life.

Lastly, though the book is based on the amazing transformation in the physical and natural sciences, in general we will try to avoid using highly technical or mathematical terms to discuss and explain this transformation – although some are unavoidable. For example, the terms 'linearity' and 'non-linearity' have very specific technical definitions within these areas and we will briefly use them. However, our intention is to try and capture the core meanings of the terms and concepts, simplify them and then build on them to explore their implications for the social sciences and public policy. Experts in the fields could easily criticise us for our lack of depth and detail when using these terms. Moreover, as argued by Edgar Morin, those who see complexity in a 'restricted' sense of only a set of new tools for analysing distinct concrete complex phenomena will be equally critical of our attempts to expand it to the social field. Nevertheless, we feel that it is necessary to simplify these concepts to reach out to social scientists and policy actors. For example, although they are interchangeable in some ways, we have chosen to use the terms 'order' and 'disorder' instead of 'linear' and 'alinear' to represent the opposite ends of complex phenomena, while 'complexity' will be used instead of 'non-linearity' to exemplify the range of phenomena and system that exist between order and disorder.

The plan of the book

The book is divided into two main parts. Part 1 briefly reviews the historical and theoretical development of the Newtonian frame of reference, rise of complexity theory within the physical sciences in the twentieth century and its increasing spill over into the social sciences in the 1980s and 1990s. After this, we outline some of the major concepts and tools of complexity that will be used in later chapters. Part 2 focuses on applying complexity theory to major political, economic and policy issues. Each of these chapters will demonstrate how the various tools of complexity (fitness landscape, complexity cascade, etc.) can be applied to that particular area. The key point to grasp is that all of the tools are relevant and can be applied to all of the policy areas. However, due to the limits of this book we will only directly explore a couple of tools in any one field.

Chapter 1 begins with a brief overview of the rise of the orderly scientific revolution in Western Europe, typified by the work of Sir Isaac Newton and Rene Descartes that laid the foundation for the Industrial Revolution and the successful development of today's most advanced industrial nations. It then explores how the orderly and linear world of Newton was challenged by the development of complexity theory in the natural sciences. Stepping through the implications of complexity for physical, biological and human systems, the chapter then builds a comparative model between orderly (modernist), complexity and disorderly (post-modernist) approaches to science and public policy.

Chapter 2 provides a short introduction to a number of major concepts of complexity theory. Moving from physical and biological to human systems, it begins with relatively simple concepts such as attractors, non-linearity and unpredictability that are found in physical complex systems. Then, it explores the processes such as punctuated equilibrium, gateway events and evolution in biotic complex systems and concludes with an examination of the dynamics of knowledge, ideas and emergence in conscious complex systems.

Chapter 3 builds on the concepts in Chapter 2 and presents six 'tools' of complexity that we will be using in later chapters. These tools are a mixture of old and new concepts related to complexity thinking. Some emerged independently of the complexity framework, soft systems methodology for example, while others come directly from what could be called an emerging complexity tradition, the 'Stacey diagram' in particular. What we plan to demonstrate is that despite its sweeping implications, complexity and its tools are actually relatively simple and easy to use. In fact, most policy actors are already using some of the tools in a 'common sense' fashion, but are unaware of the larger linkages and framework. *We believe that all of these tools are relevant and applicable to all policy areas.*

Chapter 4 begins by exploring a central paradox of national level democracy. Much of twentieth-century democratic theory focused on the supposedly clear linkage between electors, the elected and policy outcomes. Voters had preferences; these were translated to the elected who then implemented them as policies – the proverbial 'will of the people'. Unfortunately, complexity steps in and ruins this simplistic relationship. It undermines the notion that voters have clear preferences, and that those elected can deliver these preferences at will. From a complexity perspective, therefore, democracy becomes a messy incremental system for continually adapting to and adjusting the needs and desires of a complex society with evolving dynamic situations and no clear dividing line between decision-makers and stakeholders.

Likewise, the famous/infamous 'third way' concept developed by the philosopher Anthony Giddens and popularised by Tony Blair and Bill Clinton, which attempted to chart a new order beyond the market and state, will be shown to be inherently flawed and authoritarian. Despite claiming to be radically open, it falls back into the traditional trap of claiming to know the particular direction of society. From the third way, the chapter looks at why the 'third' or voluntary sector is so important to the functioning of politics and economics, particularly at the European level. To many, this sector is a messy leftover that falls between the organisational structures of the state and market. From a complexity perspective, the sector is a healthy indication of complex societal learning and interaction.

Chapter 5 explores the tremendous implications of a complexity perspective for aspects of health. Undoubtedly the orderly scientific perspective, most clearly developed by Descartes' vision of the body as a machine, formed the basis for many of the amazing advances in modern medicine and health systems. However, these advances came with a variety of costs such as a belief that health was the realm of the expert, that health systems should be fundamentally similar and that high technology and hierarchical systems were the only and best forms for health delivery. The chapter begins by exploring how public health and health systems can be reconceptualised using complexity and what this means for the promotion of global health and national health systems. Following this, the chapter drops from the global level to the individual level to explore how complexity and complexity tools can be used to help individual patients, carers and practitioners manage chronic illness.

Chapter 6 applies the concepts and ideas of complexity to the international realm. It begins by arguing that the international system, which is often seen as a simple power game or legally structured arena, should be seen as an adaptive and evolving complex system. In this massive international complex system, globalisation cannot be seen in simplistic terms such as automatic wealth creator or global economic oppressor. Globalisation becomes a subtle process of complex adaptation and evolution that shapes the global 'fitness landscape'. Similarly, for many Europhobes, Europhiles and most integration theories, the current messy, contradictory and complex state of the European Union (EU) is unhealthy and unsustainable. From a complexity perspective, the EU's messy nature is a hidden strength. Without deals, bargaining, flexible implementation, and so on, the EU would never be able to bring together all of the multiple interests that make up the European polity.

Linked to this discussion of globalisation and Europeanisation is the remarkable survival of Scandinavian exceptionalism. The small, wealthy and social democratic Scandinavian nations of Denmark, Norway and Sweden with their extensive and universal welfare states were supposed to be the victims of globalisation/Europeanisation. Their ability to adapt and thrive in their transformed surroundings provides an excellent example of the diversity of the international system and the ability of actors to find distinctive ways to adapt to it. Finally, the chapter concludes with a brief section on how complexity thinking might apply to the 2001 EU proposal to make Brussels a 'true' capital of Europe.

Chapter 7 continues the international theme by focusing on the field of development policy. The chapter begins by arguing that much of post-WWII development policy has been based on a remarkably reductionist and orderly interpretation of development and economic history and goes on to explore what a complexity view of development might look like. This includes focusing on basic social programmes, emphasising self-help and local coping strategies, promoting human rights and institutional reforms and dealing with militarism, conflict and security. The chapter concludes with a brief discussion of the reasons why a paradigm shift in development will be so hard, why it is so necessary nonetheless and how a simple change in management style (from hard to soft management) would make an immediate impact on policy outcomes.

Chapter 8 looks at the largest planning failure of the young twenty-first century, the Iraq war and its aftermath. Whether one was a supporter or opponent of the war, in planning terms it was a disaster – costing over one trillion dollars and disrupting millions of lives. In this chapter we argue that despite a major planning effort and an easy military victory, winning the peace and significantly altering Iraqi society (a remarkably complicated and deeply historical complex system) in a short period of time was doomed to failure. More money, planning, effort or different leaders would have been unlikely to make a significant difference since the fundamental approach to restructuring Iraq was based on orderly premises defined and imposed in an unusually strict top-down manner.

Chapter 9 continues to explore the costs of using orderly frameworks to direct a complex, unpredictable world by examining the myths that have grown up around the issue of terrorism. From a traditional orderly perspective these appear to be reasonable assumptions that are currently guiding much national and international foreign policy. On the other hand, from a complexity perspective, they are remarkably misguided assumptions that misdirect and undermine sensible foreign policies. In essence, to achieve an orderly victory in the 'war on terror' the leading powers not only will have to devote massive and continuing effort to the task but in doing so their societies will have to undergo radical transformations that would undo internal achievements (civil liberties, legal rights, prohibitions against torture, etc.) that were accomplished after centuries of hard efforts. Both the Obama administration and Brown government have shown signs of learning some of these lessons, but not grasping the full systemic implications.

Chapter 10, in conclusion, will briefly respond to some of the main questions that students and policy actors have confronted us with over the past ten years and finish with a few playful mental exercises. In doing this we argue that complexity will not replace the Newtonian paradigm, but will build upon its orderly foundation. Hopefully, the book will generate a better awareness of when to use orderly and complexity methods and help to avoid the extremes of order and disorder in the twenty-first century.

1 From orderly to complexity science

> The end of our foundation is the knowledge of causes, and secret motions of things: and the enlarging of the bounds of human Empire, to the effecting of all things possible.
>
> (Francis Bacon, *The New Atlantis* 1627)

What is the complexity paradigm? How and when did it emerge? Is it a hot new academic fad like globalisation or the end of history, or is it something more profound? To begin to answer these questions we need to jump back a few centuries and briefly discuss the emergence of what is commonly labelled the 'Newtonian' or 'linear' paradigm. For reasons that will become clear, we have called it 'the paradigm of order'.

The paradigm of order

Although it has been said thousands of times before, it bears repeating that the Enlightenment was an astounding time for Europe. Relatively stagnant and weak and intellectually repressed by the Church during the so-called Dark Ages, intellectual energies released by the Renaissance came to fruition in the Enlightenment. During this time, Europe was reborn and became the centre of an intellectual, technical and economic transformation. It had an enormous impact on the way life is viewed at all levels from the mundane to the profound. Science was liberated from centuries of control by religious stipulations and blind trust in ancient philosophies. Rene Descartes (1596–1650) and, slightly later, Sir Isaac Newton (1642–1727) set the scene. The former advocated rationalism while the latter unearthed a wondrous collection of fundamental physical laws. A flood of other discoveries in diverse fields such as magnetism, electricity, astronomy and chemistry soon followed, injecting a heightened sense of confidence in the power of reason to tackle any situation. The growing sense of human achievement led the famous author and scientist Alexander Pope to poeticise, 'Nature, and Nature's laws lay hid in night. God said *Let Newton be!* And all was light'.[1] Later, the eighteenth century French scientist and author of *Celestial Mechanics*,

Pierre Simon de Laplace (1749–1827), carried the underlying determinism of the Newtonian framework to its logical conclusion by arguing that 'if at one time, we knew the positions and motion of all the particles in the universe, then we could calculate their behaviour at any other time, in the past or future.'

The subsequent phenomenal success of the industrial revolution in the eighteenth and nineteenth centuries, which was based on this new scientific approach, heightened confidence in the power of human reason to tackle any physical situation. By the late nineteenth and early twentieth century many scientists believed that few surprises remained to be discovered. For the American Nobel Laureate, Albert Michelson (1852–1931), 'the future truths of Physical Science are to be looked for in the sixth place of decimals' (Horgan 1996: 19),[2] implying that physicists were now only filling in the small cracks in human knowledge. More fundamentally, the assumption and expectation was that over time the orderly nature of all phenomena would eventually be revealed to the human mind. Science became the search for hidden order. The universe and everything in it is a magnificent clockwork mechanism.

By and large, that vision of the universe survived well into the twentieth century. In 1996 John Horgan, a senior writer at *Scientific American*, published a bestselling book entitled *The End of Science* which argued that since science was linear and all the major discoveries had been made, then real science had come to an end. All that was left was 'ironic science' which 'does not make any significant contributions to knowledge itself. Ironic science is thus less akin to science in the traditional sense than to literary criticism – or to philosophy' (Horgan 1996: 31).

Similarly, the eminent biologist and Pulitzer prize winner, Edward O. Wilson argued in his bestselling book *Consilience* (1998) that all science should be unified in a fundamentally linear framework based on physics:

> The central idea of the consilience world view is that all tangible phenomena, from the birth of stars to the workings of social institutions, are based on material processes that are ultimately reducible, however long and tortuous the sequences, to the laws of physics.
>
> (Wilson 1998: 291)

The orderly view of the world prospered not only in sciences, but in the fundamental nature of Western social and political life.

To simplify drastically, the paradigm of order was founded on four golden rules:

- *Order*: given causes lead to known effects at all times and places.
- *Reductionism*: the behaviour of a system could be understood, clockwork fashion, by observing the behaviour of its parts. There are no hidden surprises; the whole is the sum of the parts, no more and no less.
- *Predictability*: once global behaviour is defined, the future course of events could be predicted by application of the appropriate inputs to the model.
- *Determinism*: processes flow along orderly and predictable paths that have clear beginnings and rational ends.

14 *From orderly to complexity science*

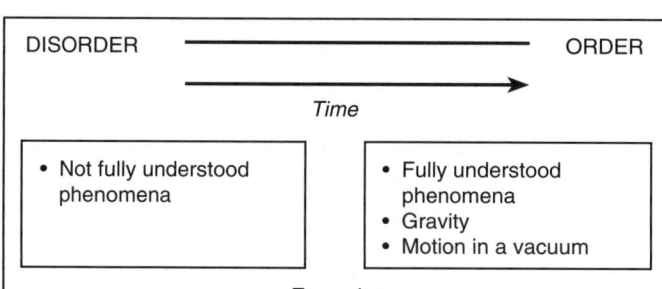

Figure 1.1 Phenomena in the paradigm of order.

From these golden rules a simple picture of reality emerged.

- Over time as human knowledge increases, phenomena will shift from the disorderly to the orderly side.
- Knowledge equals order. Hence, greater knowledge equals greater order.
- With greater knowledge/order humans can increasingly predict and control more and more phenomena, including human phenomena and systems.
- There is an endpoint to phenomena and hence knowledge.

The orderly paradigm worked remarkably well and was conspicuous by incredible leaps in technological, scientific and industrial achievements. Science became orderly and hierarchical with clear divisions that manifested themselves in the departmentalised evolution of modern universities. Not surprisingly, success in these areas had a profound effect on attitudes in all sectors of human activity, spreading well beyond the disciplines covered by the original discoveries.

Spreading ripples of doubt

Certainty and predictability for all, the hallmarks of an orderly frame of mind, were too good to last. Fissures had existed for some time, even Isaac Newton and Christiaan Huygens in the seventeenth century couldn't agree on something as fundamental as the nature of light (whether it is a particle or a wave). These difficulties bubbled under the surface of acceptable scientific discourse and the expanding university arenas. They were often seen as unimportant phenomena that would be resolved by the next wave of emerging fundamental laws. However, by the early twentieth century they could no longer be ignored. The physicist Henri Poincaré (1854–1912) was one of the first to voice disquiet about some contemporary scientific beliefs. He advanced ideas that predated chaos theory by some seventy years (Coveney and Highfield 1995: 169). Later, Einstein's (1879–1955) theory of relativity, Neils Bohr's (1885–1962) contribution to quantum mechanics, Erwin Schrödinger's (1887–1961) quantum measurement problem, Werner Heisenberg's (1901–76) Uncertainty Principle and Paul A.M. Dirac's (1902–84) work on quantum field theory all played a decisive role in pushing conventional wisdom beyond the Newtonian limits that had enclosed it centuries before. These scientists, all Nobel Laureates, set in motion a process that eventually transformed attitudes in many other disciplines.[3]

The new discoveries did not disprove Newton. Essentially, they revealed that not all phenomena were orderly, reducible, predictable and/or deterministic. For example, no matter how hard classical physicists tried they could not fit the dualistic nature of light as both a wave and a particle into the orderly classical system. Heisenberg's Uncertainty Principle, which shows that one can either know the momentum or position of a sub-atomic particle, but not both at the same time, presents an obvious problem for the orderly paradigm. Or, the paradox of Schrödinger's Cat experiment, which demonstrated the distinctive nature of quantum probability and again broke the fundamental boundaries of the former order. What this meant was that even at the most fundamental level some phenomena do conform to the classical framework and others do not. With this, the boundaries of the classical paradigm were cast asunder. Gravity continued to function and linear mechanics continued to work, but it could no longer claim to be universally applicable to all physical phenomena. It had to live alongside phenomena and theories that were essentially *probabilistic*. They did not conform to the four golden rules associated with linearity: order, reductionism, predictability and determinism. Causes and effects are not linked, the whole is not simply the sum of the parts; *emergent properties* often appear seemingly out of the blue, taking the system apart does not reveal much about its global behaviour, and the related processes do not steer the systems to inevitable and distinct ends.

Given these non-linear phenomena and non-adherence to the golden rules of order, new expectations were necessary for this expanding paradigm:

- Over time human knowledge may increase, but phenomena will not necessarily shift from the disorderly to the orderly.
- Knowledge does not always equal order. Greater knowledge may mean the increasing recognition of the limits of order/knowledge.

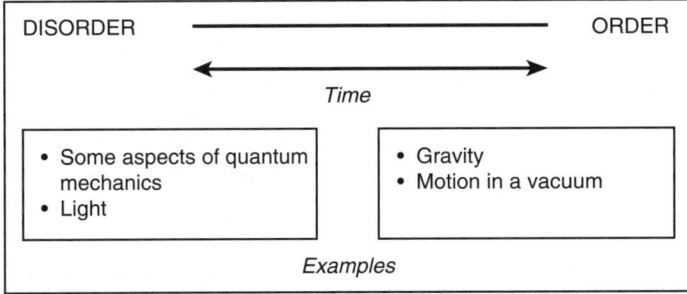

Figure 1.2 Phenomena in the paradigm disorder and order.

- Greater knowledge does not necessarily impart greater prediction and control. Greater knowledge may indicate increasing limitations to prediction and control.
- There is no universal structure/endpoint to phenomena/knowledge.

It is important to note that the shift in scientific analysis from utter certainty to considerations of probability was not accepted lightly. Schrödinger had originally designed his cat experiment as a way of eliminating the duality problem! The sea change radiated slowly outwards from quantum mechanics' domain of sub-atomic particles. Naturally, there was a wide schism between the exclusive niches occupied by leading particle physicists and mathematicians on the one hand, and the rest of the scientific community on the other. High specialisation meant that even scholars involved in the same discipline were not immediately aware of discoveries being made by their colleagues. Moreover, the language of science itself became almost unintelligible beyond a select circle of specialists. In any case, their intriguing speculations were not thought at first to be of everyday concern. Nevertheless, uncertainty was eventually recognised as an inevitable feature of some situations. In effect, the envelope of orderly science was expanded to embrace complex phenomena, also known as *complex systems*, to those already in place.

Complex systems in the physical world

Once the door was open to probability and uncertainty, a new wave of scientists began studying phenomena that had previously been ignored or considered secondary or uninteresting, aspects that the Nobel prize winner Ernst Rutherford disparagingly called 'stamp-collecting' activities (Briks 1962: 108).[4] Weather patterns, fluid dynamics and Boolean networks were just three of the areas that saw the growing acceptance of non-linear complex phenomena and systems. For example, one of the earliest people to conceptualise and model a complex system was an American meteorologist, Edward Lorenz (see Gleick 1987). Lorenz developed a computer programme for modelling weather systems in 1961. However, to his dismay due to a slight discrepancy in his initial programme, the programme produced wildly divergent patterns. How was this possible? From an orderly framework, small differences in initial conditions should only lead to small differences in outcomes. But, in Lorenz's programme, small discrepancies experienced positive feedback and reinforced themselves in chaotic ways producing radically divergent outcomes. Lorenz called this the 'butterfly effect', arguing that given the appropriate circumstances a butterfly flapping its wings in China could eventually lead to a tornado in the USA. Cause did not lead to effect. Order was not certain. Chaos/complexity was an integral part of physical phenomena. Moreover, some phenomena could not be reduced and isolated, but had to be seen as part of larger systems.

Other examples of complex systems can be found in simple forms of fluid dynamics. For example, the water molecules creating a vortex in your bathtub is a type of physical complex system. The molecules self-organise and form a stable complex system so long as the water lasts in the bathtub. The vortex is easy to recreate, but the exact combination of water molecules that made the specific vortex would be virtually impossible to reconstruct. Each vortex, though similar, is not an exact copy of the other. The Lorenzian Waterwheel, a simple waterwheel that exhibits orderly and chaotic motion merely by manipulating the flow of water, is another classic case (Gleick 1987: 27).

This systems approach led to the creation of a variety of definitions of complex systems. In the physical world these systems are described as being *complex*, because they have numerous internal elements; *dynamic*, because their global behaviour is governed by local interactions between the elements; and *dissipative*, because they have to consume energy to maintain stable global patterns. Physical complex systems obey fundamental physical laws, but not in the same way as orderly systems. For example, the second law of thermodynamics, the most fundamental law of nature, states that when a system is left alone it drifts steadily into disorder. The effects of the second law are plain to see. A deserted building, for instance, eventually turns into a pile of rubble. After a few centuries even the rubble disappears without a trace. Ultimately, a system cut off from the outside world will fall into a deathly state of equilibrium in which change does not occur. For the complexity physicist Peter Allen, orderly equilibrium systems are 'dead' systems (Allen 2001).

Orderly systems are found at or near equilibrium. A ball bearing inside a bowl is a classic example; it quickly settles at the bottom and that is that. These systems can be very complicated. A jet engine is a wonderfully complicated piece of orderly machinery creating highly predictable physical outcomes that millions of pilots and passengers successfully depend upon every year. Complexity, by contrast, is exhibited by systems that are far from equilibrium. In this instance, the system has to exchange (dissipate) energy, or matter, with other systems in order to acquire and maintain self-organised stable patterns. That is the only option open to it to avoid falling into the destructive clutches of the second law of thermodynamics. The most dramatic illustration of that process is planet Earth. Without the nourishing rays of energy from the Sun, Earth would perish into complete equilibrium, and therefore nothingness. Continuous supply of energy from the Sun keeps the planet in a highly active state far from equilibrium. The energy is absorbed, dissipated and used to drive numerous local interactions that in total produce the stable pattern that we perceive as life on Earth.[5]

The golden rules for physical systems in a complexity paradigm are as follows:

- *Partial order*: phenomena can exhibit both orderly and chaotic behaviours.
- *Reductionism and holism*: some phenomena are reducible others are not.
- *Predictability and uncertainty*: phenomena can be partially modelled, predicted and controlled.
- *Probabilistic*: there are general boundaries to most phenomena, but within these boundaries exact outcomes are uncertain.

18 *From orderly to complexity science*

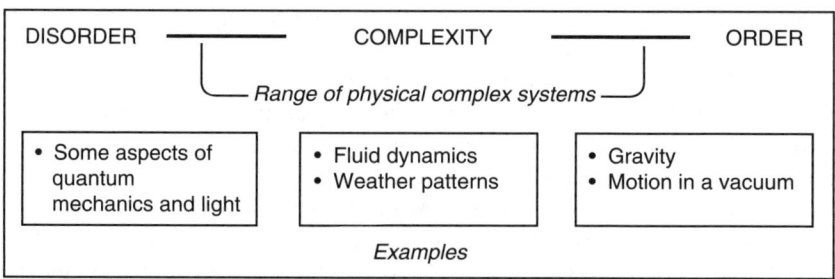

Figure 1.3 The range of physical phenomena in a complexity paradigm.

Complex systems in the biotic world

By the latter half of the twentieth century, with complexity already deeply penetrating the physical sciences, biologists, geneticists, environmentalists and physiologists also began to consider their respective disciplines within the context of complexity.[6] Analysts in these fields set out to investigate the properties of systems, including human beings, comprised of a large number of internal parts that interacted locally in what looked like a state of anarchy that somehow managed to engender self-organised, stable and sustainable global order. These systems were not only complex, dynamic and dissipative, but also adaptive and display *emergent properties* or *emergence*.

In the words of Murray Gell-Man, a Nobel prize-winning physicist, 'turbulent flow in a liquid is a complex system ... But it doesn't produce a schema, a compression of information with which it can predict the environment' (Lewin 1999: 15). Without that schema, non-biological systems cannot respond to their environments in anything other than orderly, disorderly or physically complex ways. The ability of biotic complex systems to adapt and evolve creates a whole new range of complex outcomes. Likewise, biological complex systems are able to develop new emergent properties that may reshape the complex system as a whole and/or the sub-units that make up the system. As Coveney and Highfield argue: 'Life is also an emergent property, one that arises when physiochemical systems are organized and interact in certain ways' (Coveney and Highfield 1995: 330).

From this perspective a whole new range of biotic complex systems began to be studied. For example, S. Kauffman (1993) was one of the first to view the genetic code as an evolving complex system. Other concepts like autopoiesis, symbiosis and the *Gaia* system emerged to challenge the orderly framework in the biological sphere (Capra 1996; Lovelock 1979; Margolis 1993). Due to the emergent nature of biological systems, the level of complexity can be significantly higher than those of physical phenomena and systems. Hence, on our simple scale of complexity biotic complexity is placed on the more disorderly side of the scale than biotic complexity.

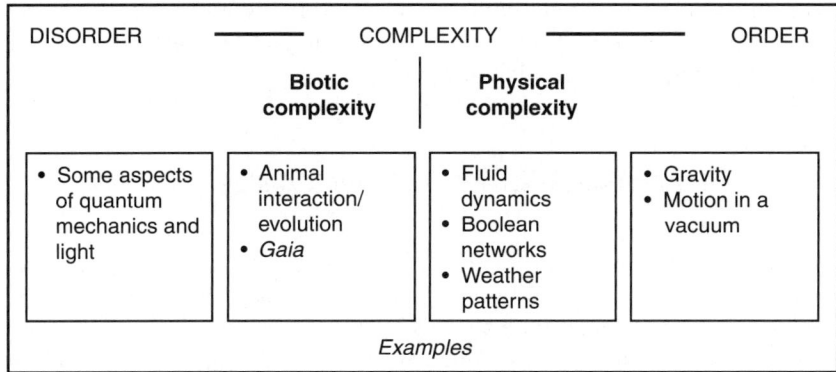

Figure 1.4 The range of physical and biotic phenomena.

The golden rules of biotic systems in a complexity paradigm are as follows:

- *Partial order*: phenomena can exhibit both orderly and chaotic behaviours.
- *Reductionism and holism*: some phenomena are reducible others are not.
- *Predictability and uncertainty*: phenomena can be partially modelled, predicted and controlled.
- *Probabilistic*: there are general boundaries to most phenomena, but within these boundaries exact outcomes are uncertain.
- *Emergence*: they exhibit elements of adaptation and emergence.

A simple example of a biotic complex system would be the evolution of a species or the interaction of a given plant or animal in a particular ecosystem. A fish in a small pond will evolve and interact with the various food sources (small plants and animals) in the pond to create a stable complex system (such as a stable total number of fish). However, if a change is introduced to the system, a new competitor or food source, the fish may adapt and alter the nature of the system in totally unforeseen ways. Over time, new emergent properties may evolve in the system and/or in the fish itself.

A larger example is that of the concept of *Gaia*. As summarised in Coveney and Highfield:

> In 1968, James Lovelock upset gene-centered proponents of Darwin's views by arguing that the earth was not a ball of rock with a green layer of life on the surface. Biologists, following Darwin, see life adapting to its environment. The independently minded Lovelock viewed life and the environment as part of one superorganism in which creatures, rocks, air, and water interact in subtle ways to ensure that the environment remains stable ... feedback mechanisms are invoked to explain the relative constancy of the climate, the

surprisingly moderate levels of salt in the oceans, the constant level of oxygen over the past few hundred million years, and why life forms are so diverse. Like it or hate it, simply looking for Gaia can give new insights into the complex feedback systems that rule the planet.

(Coveney and Highfield 1995: 234–35)

Orderly (modernist) social science and public policy

The success of the orderly paradigm in the natural sciences had a profound effect on attitudes and practices in all sectors of human activity. The social sciences were no exception. Surrounded by the technological marvels of the industrial revolution which were founded on a Newtonian vision of an orderly, clockwork universe driven by observable and immutable laws, it did not take much of an intellectual leap to apply the lessons of the physical sciences to the social realm. The English philosopher Thomas Hobbes (1588–1678) used a mechanistic vision to shape an orderly society, a *Leviathan* that would save humans from chaos and civil war. The French economist Francois Quesnay (1694–1774) and the *physiocrates* modelled the economic system on a mechanical clock. The French mathematician, philosopher and revolutionary politician Condorcet (1743–94) wrote while imprisoned by the Committee of Public Safety:

> The sole foundation for belief in the natural sciences is the idea that the general laws directing the phenomena of the universe, known or unknown, are necessary and constant. Why should this principle be any less true for the development of the intellectual and moral faculties of man than for other operations of nature?
>
> (Wilson 1998: 21)

The famous British economist Adam Smith (1723–90) claimed to have captured the laws of economic interaction while his follower, David Ricardo (1772–1823) believed that some economic laws were 'as certain as the principles of gravitation' (Mainzer 1997: 264). Karl Marx (1818–83) wedded his vision of class struggle to an analysis of the capitalist mode of production to create the 'immutable' and deterministic laws of capitalist development. Academics in all the major fields of social science welcomed the new age of certainty and predictability with open arms. Economics, politics, sociology all became 'sciences', desperate to duplicate the success of the natural sciences. Moreover, this desire was institutionalised through the development of modern universities that created and reinforced the disciplinarisation and professionalisation of the social sciences (Gulbenkian Commission 1996: 7).

The high point of the orderly paradigm was reached in the 1950s and 1960s, particularly in universities in the United States. Strengthened by the success of planning programmes during WWII and the early post-war period, pressured by the growing Cold War, and lavishly funded by the expanding universities, American academics strived to demonstrate, and hence control, the presumed rational nature of human interaction. This traditional Newtonian approach was clearly expressed in the modernisation theories of Third World development, the realist vision of international relations, the behaviouralist writings of sociologists, the positivist foundations of liberal economics and the rational plans of public-policy experts and social planners.

Using the Newtonian frame of reference modern social scientists unjustifiably assumed that physical and social phenomena were primarily orderly and therefore predictable. They consequently applied reductionist methods founded on the belief that stable relationships exist between causes and effects, such as the assumption that individual self-interest is an explanation and/or a model for national-level self-interest. Furthermore, based on this orderly thinking they assumed that society and social institutions had an 'end-state' towards which they were evolving. Hence, economic interaction, democracy, fundamental social orders (communism, capitalism, development) and so on, all had final stages towards which they were evolving. Nation-states, societies and even individuals could be positioned along this developmental pathway and policies could be devised to help them towards the next level. As Table 1.1 summarises, orderly social science rested on the same foundation as orderly natural science, treated human beings like orderly atomistic objects and drew similar orderly conclusions.

Unsurprisingly, public policy, resting on a foundation of social science, was equally dominated by the Newtonian vision. In one of the classics in the field, *Seeing Like a State*, the distinguished Yale University professor James Scott categorised a lengthy nineteenth and twentieth century history of large scale public-policy blunders by democratic, authoritarian, advanced and developing states. Ranging from plans for mapping cities, resources and populations in Europe and the USA to the disasters of agricultural collectivisation in the USSR, villagisation in Tanzania and the designed cities in Brasil, Scott notes that:

> Radically simplified designs for social organization seem to court the same risks of failure courted by radically simplified designs for natural environments. The failures and vulnerability of monocrop commercial forests and genetically engineered, mechanized monocropping mimic the failures of collective farms and planned cities
>
> (Scott 1998: 7)

Scott, not mentioning complexity but having very similar ideas, goes on to note that four factors were essential for the creation of the most grandiose public-policy failures. These were: (1) the normal administrative and bureaucratic tools of large scale state institutions, (2) a belief in what Scott calls, 'high modernist ideology' that views the 'rational design of social order commensurate with the scientific understanding of natural laws' (Scott 1998: 4), (3) an authoritarian state, and (4) a weak civil society unable to resist the state's plans.[7]

In the case of twentieth-century public policy one can scan through a variety of public-policy textbooks to review the predominately orderly/modernist evolution of public policy in the West. Most begin with the illustrious work of Max Weber,

Table 1.1 The foundations of orderly (modernist) social science

Theoretical basis

Order, reductionism, predictability and determinism

Expectations

- Over time human knowledge increases, phenomena will shift from the disorderly to the orderly realm. *Social scientists are able to understand more and more about society and humanity.*
- Knowledge equals order. Hence, greater knowledge equals greater order. *Thus, history is progressive, leading to greater order.*
- With greater knowledge humans can increasingly predict and control more and more phenomena. *Those with greater knowledge can know more and thus should be in control.*
- There is an equilibrium or endpoint to all phenomena. *Once this endpoint is reached history stops and societal change comes to an end.*
- There is a hierarchy of scientific knowledge and methods with the orderly natural sciences at the zenith. *Duplicating this knowledge and methods is the primary justification of orderly social science.*

Implications

- Researchers look for rational foundations to all phenomena.
- There are no limits to human knowledge. The only constraints are effort and technology.
- Researchers can obtain predictable, verifiable, repeatable experimental results.
- Duplicating orderly natural science methods is the primary methodological strategy.
- The creation of universal, stable and parsimonious social laws is the ultimate goal.

Vision of history and progress

- Over time, social scientists can discover the laws of the human realm and move societies from disorder to order, progressing them towards a final ultimate order.

later summarised by Talcott Parsons, who mapped out the early contours of the modern state. These included:

- Centralised pyramidal administrative hierarchies
- Clears lines of responsibility
- Hierarchical discipline
- Responsibility and decision-making concentrated at the top
- Command and control procedures
- Organisational rigidity
- Uniformity and predictability
- And an ideology of meritocratic technological rationalism

The so-called British Westminster model is the ultimate example of the type of modernist-thinking public policy (Richards and Smith 2002). Demonstrating all of the above factors and more, the British state structures in Whitehall could reasonably be seen to be the model of orderly modernist administration. Presided over by a stable parliamentary system that had only two major parties, many saw the Westminster model as a major achievement in social order and stability.

Obviously, other forms of administrative structures abounded in different countries and the Westminster model itself changed over the course of the twentieth century. However, its core orderly and modernist elements remained and continued to dominate the functioning of the British state throughout the twentieth century. Interestingly, as we will explore in later chapters, when challenges and problems occurred within the Westminster model it often reverted to a more intensified order or modernism. For example, during the 1970s the British state was often seen as being 'overloaded' with welfarist, social and economic demands. Margaret Thatcher rode to power on the crest of this criticism and the belief that a greater reliance on market mechanisms would solve many of Britain's ills. However, after arriving in power she found that she had to concentrate and focus the powers of the central state in order to control the disparate actions of local actors.

In terms of public-policy thinking the 1980s and 1990s were epitomised by the rise of public choice thinking and new public management. Public choice thinking, founded on the work of the American Antony Downs (1957), was an attempt to bring the rationality and order of neo-classical economics to public policy and public sector actors. For many public choice theorists, public bureaucrats needed to be viewed in the same way as the idealised economic man was in mainstream economics, a rational utility-maximising actor that was not interested in the public good but in the maximisation of their own bureaucratic interests. The problem was that without some form of market to constrain these rational bureaucratic actors they would just keep on demanding more and more resources to increase their ever-expanding administrative empires. For right-wing thinkers in the 1980s public choice thinking provided an explanation for the general rise in state expenditure and public bureaucracies during the twentieth century and the means, more markets (implying the sale of state assets and state monopolies) and market-like constraints, for countering the tendency of bureaucratic expansion. However, despite great efforts large state structures and public support for them refused to go away while at the same time integrating markets into state activities produced their own variety of malfunctions.

By the 1990s, public choice theory had metamorphosed into New Public Management (Hood 1991). NPM thinkers wedded the belief in the superiority of the market with a desire to decentralise and increase the flexibility of public administration and policy. However, to justify this decentralisation and flexibility required a radical increase in performance indictors so that the centre could oversee and direct what the local and decentralised actors were doing (Walsh 1995). This focus on efficiency and outputs led to an explosion of public-policy targets and the growth of the 'audit society' (Power 1997), embraced by both Left and Right. Once again, attempts at creating greater flexibility and variety in the administration and outputs of public policy were undermined by an overriding desire for central control and oversight. Hence, despite huge efforts and a clear understanding of the weaknesses of the traditional orderly Westminster model of public policy, public choice and NPM couldn't go beyond the existing confines of the traditional framework.

The basic outlines of the orderly/modernist vision of public policy can be outline in Table 1.2.

Disorderly (post-modernist) social science and public policy

Even at its peak, countervailing tendencies in the social sciences survived. There is nothing new about questioning the fundamental order and rationality

Table 1.2 The foundations of orderly (modernist) public policy

Theoretical basis

Order, Reductionism, Predictability and Determinism

Expectations

- Over time human knowledge increases, phenomena will shift from the disorderly to the orderly realm. *Central policy actors are able to understand more and more about their societies and humanity in general.*
- Knowledge equals order. Hence, greater knowledge equals greater order. *More knowledgeable 'evidence-based' policies will create more order.*
- With greater knowledge humans can increasingly predict and control more and more phenomena. *Policy elites with greater knowledge can know more and thus should be in control.*
- There is an equilibrium or endpoint to all phenomena. *All public policy should strive to reach this point. When it is reached all fundamental policy/societal change will stop.*
- There is a hierarchy of scientific knowledge and methods with the orderly natural sciences at the zenith. *Duplicating this knowledge and methods is the primary justification of orderly public policy.*

Implications

- Public policy actors must look for and act on rational foundations in all public policy areas.
- There are no limits to human knowledge and public policy. The only constraints are effort and technology.
- Public policy actors can obtain predictable, verifiable and repeatable policy outcomes.
- Duplicating orderly natural science methods is the 'gold standard' of public policy methods.
- The creation of an improved and stable order is the ultimate goal.

Vision of history and progress

- Over time, with appropriate policies and power, policy actors can move societies from disorder to order and progress them towards a final ultimate order.

of human existence. Debates over theses issues are easily traced back to Plato and Aristotle.[8] A belief in the fundamentally rational and orderly nature of human existence only emerged in the Western philosophical tradition in the seventeenth and eighteenth centuries. Before this period, much of the human and physical world embraced unknowable mysteries that were cloaked in the enigmas of religion. During the eighteenth, nineteenth and twentieth centuries, there continued to be a huge variety of potent critics of the mechanistic view of nature and society and of the limits of human rationality. In the late eighteenth century, the German scientist and philosopher, Immanuel Kant (1724–1804) argued that an organism, 'cannot only be a machine, because a machine has only moving force; but an organism has an organising force ... which cannot be explained by mechanical motion alone' (Mainzer 1997: 83).[9]

These arguments plus the work of Friedrich Schelling (1775–1854) who described an organic 'science of living' and the writings of Goethe (1749–1832) who saw the mechanistic model of nature as 'grey ... like death ... a ghost and without sun' (Mainzer 1997: 84) created the foundation of the German romantic philosophy of nature which rejected the mechanism of Newton. In the early twentieth century, the hermeneutical tradition of Sigmund Freud (1865–1939) and Max Weber (1864–1920) challenged the belief in the human rational capabilities and the degree to which humans can understand and control their environment and societies.

In the mid-twentieth century, the American philosopher John Dewey (1859–1952) was advocating his philosophy of pragmatism as a strategy for dealing with the limits of knowledge and uniqueness of human experience while the German sociologists Theodor Adorno and Max Horkheimer were publishing their masterwork, *Dialectic of Enlightenment*, that critiqued the Enlightenment foundations of modernism. In the 1960s the famous Austrian economist F.A. Hayek (1899–1992) argued that: 'In the field of complex phenomena the term "law" as well as the concepts of cause and effect are not applicable' (Hayek 1967: 42). By the 1970s, the influential French post-modernist philosopher Jean-Francois Lyotard, in *The Postmodern Condition: A Report on Knowledge* (1984) was arguing for an end to all 'grand narratives' of Western society. Consequently, from the 1970s onwards as social scientists continually failed to capture the 'laws'[10] of society and economic interaction and were continually frustrated over their inability to do so, they began to significantly question the Newtonian framework that underpinned political thinking on the Left and Right.[11]

Out of this emerged the extremely diverse, but significant challenge of (disorderly) post-modern position in social science. As defined by Terry Eagleton:

> Postmodernism ... is a style of thought which is suspicious of classical notions of truth, reason, identity and objectivity, of the idea of universal progress or emancipation, of single frameworks, grand narratives or ultimate grounds of explanation. Against these Enlightenment norms, it sees the world as contingent, ungrounded, diverse, unstable, indeterminate, a set of disunified cultures or interpretations which breed a degree of scepticism about the

objectivity of truth, history and norms, the givenness of natures and the coherence of identities.

(Eagleton 1996: vii)

As excellently summarised by Colin Hay (2002: 227), the post-modernist position stands in direct contrast to the traditional orderly (modernist) social science position.

Mirroring these post-modernist criticisms in social science, there has been a wide range of twentieth-century challengers to the dominance of the orderly public-policy framework. In the early twentieth century thinkers like Max Weber, though he is well-known for being among the first to describe and analyse traditional administrative and organisational structures, were noting how organisational and institutional structures were capable of altering the underlying rationality of public-policy actors to produce misguided and inappropriate outcomes, the so-called 'iron cage' of rule based rational control. In his book,

Table 1.3 The foundations of disorderly (post-modern) social science

Theoretical basis

- Reality and rationality are relational and experienced differently depending on specific cultural and temporal dynamics.
- Reality is unpredictable, irreducible and indeterminate.

Expectations

- Knowledge does not progress and is always contested.
- Knowledge is based on perspective and different perspectives are incommensurate.
- Truth claims cannot be adjudicated empirically.
- All truth claims are dogmatic and potentially totalitarian.

Strategies

- Undermine strong knowledge claims.
- Undermine modernist assumptions of privileged knowledge and power.
- Use 'deconstructivist' techniques to disrupt modernist meta-narratives and draw attention to otherwise marginalised 'others'.

Relationship between natural and social sciences

- No clear relationship exists. Relational and interpretive nature of human condition makes clear relationship impossible.

Vision of history and progress

- Multiple contested relational 'orders' flowing over time. No pure knowledge creation or progress nor direction to history.

Political Parties (1911), Robert Michels developed his 'iron law of oligarchy' arguing that all organisations become more and more oligarchical as they become larger and more complex. Hence, public-policy actors could never directly represent society. The American philosopher John Dewey using his 'pragmatist' philosophical approach argued in the 1930s for increasing flexibility and adaptability in US education policy, moving away from centrally determined educational routines that were common at the time. In the 1950s a number of public-policy thinkers began to espouse the theory of incrementalism (Lindblom 1959). This theory argued that due to the growing complexity of modern states and the fundamental uncertainty of public-policy actions, one should avoid rigid and detailed large-scale plans in favour of evolving incremental actions that can be continuously reviewed and evaluated.

Building on this, Graham Allison, an esteemed Harvard professor, argued that in any major public-policy or governmental decision there were three competing and interacting processes at work: rational actions, organisational processes and political dynamics. The recognition of the limits of rational behaviour in large organisations and public administration became so clear that in 1972 Michael Cohen, James March and Johan Olsen postulated that many public decisions were based on a 'garbage can model'[12] of decision-making; confronted with complicated and uncertain problems, policy actors reached into a garbage can of traditional strategies and applied them almost randomly to see if they would work. During the 1980s and 1990s a whole range of political scientists and policy academics began conceiving of public policy as much more of a soft and indirect form of 'governance' rather than firm 'government'. This was particularly embraced by academics studying the messy and contradictory actions and policies of the European Union who then coined the concept of 'multi-level governance' (Marks *et al.* 1996). In addition to all of these criticisms, a whole range of more radical critiques emerged from gender and race-oriented authors arguing that traditional public policy was fundamentally misshapen and distorted by society's gender and racial biases and characteristics. Obviously, many of these individuals would not class themselves as postmodernists or disorderly thinkers. However, given their underlying tendencies one can create an idealised vision of a disorderly/post-modern framework of public policy.

Following this review of the orderly and disorderly perspectives, it is important to grasp three key points. First, these summaries are obviously brief and idealised. Many of the authors would reject being placed in the given category and most of them interwove a number of different perspectives. Nevertheless, these were two clear tendencies of the twentieth century. Second, one should notice their oppositional nature. The logic of each side drives one towards equilibrium and order on one side or relativism and disorder on the other. Not that all of the authors went that far, but the tendency remains. Third, despite their incompatible natures they both clearly represent elements of reality, but not the whole picture. Is there any way to clearly and parsimoniously bridge this gap?

28 *From orderly to complexity science*

Table 1.4 The foundations of disorderly (post-modern) public policy

Theoretical basis

- Reality and rationality are relational and experienced differently depending on specific cultural and temporal dynamics.
- Reality is unpredictable, irreducible and indeterminate.

Expectations

- Knowledge does not progress and is always contested. *All policies are contested.*
- Knowledge is based on perspective and different perspectives are incommensurate. *All policies can be viewed from different perspectives and many of these perspectives are irreconcilable.*
- Truth claims cannot be adjudicated empirically. *There is no way of knowing the appropriate policy.*
- All truth claims are dogmatic and potentially totalitarian. *Policy certainty is an indicator of rigid thinking and potential oppression.*

Strategies

- Undermine strong knowledge claims. *Oppose modernist policy claims.*
- Undermine modernist assumptions of privileged knowledge and power. *Undermine modernist policy assumptions.*
- Use 'deconstructivist' techniques to disrupt modernist meta-narratives and draw attention to otherwise marginalised 'others'. *Deconstruct modernist policies and recognise the role of the 'other'.*

Relationship between natural and social sciences

- No clear relationship exists. Relational and interpretive nature of human condition makes clear relationship impossible. The *natural sciences cannot be used as a foundation for policy ideas.*

Vision of history and progress

- Multiple contested relational 'orders' flowing over time. No pure knowledge creation or progress nor direction to history. *Policy merely responds to relational flows. No foundation upon which to base progress or direction to history.*

Complexity and social science

To answer that question we have to find out how human beings fit into the complexity paradigm? They are an obvious symbiotic part of the complex web of their physical and biological surroundings. Nevertheless, what makes them distinct from this environment? Their most fundamental difference is consciousness: the ability to ask 'Who am I?', 'How did I get here?', 'What does life mean?'.

This ability to be self-aware, to understand aspects of the world around them, be aware of their history and to evolve interpretations of themselves, their surroundings and their history makes human beings fundamentally different from all other life forms and physical phenomena. However, this interpretive ability does not produce orderly interpretations. The uniqueness of individual human experience combined with multitudinous possibilities of collective human interaction and the evolutionary nature of human society produce a very high degree of complex interpretive outcomes. Therefore, conscious interpretive outcomes (norms, values, historical interpretation) must be positioned on the more disorderly side of our complexity scale. This does not imply that there are no universal norms, values or interpretations. For example, a prohibition against murder is a common societal trait. However, the definition of murder, the mitigating circumstances which could surround it and the punishment for the act all vary widely over time and between different societies and cultures. The position of conscious phenomena is outlined in Figure 1.5.

The golden rules of conscious systems in a complexity paradigm are as follows:

- *Partial order*: phenomena can exhibit both orderly and chaotic behaviours.
- *Reductionism and holism*: some phenomena are reducible others are not.
- *Predictability and uncertainty*: phenomena can be partially modelled, predicted and controlled.
- *Probabilistic*: there are general boundaries to most phenomena, but within these boundaries exact outcomes are uncertain.
- *Emergence*: they exhibit elements of adaptation and emergence.
- *Interpretation*: the actors in the system can be aware of themselves, the system and their history and may strive to interpret and direct themselves and the system.

In essence, complexity theory does not disprove the rationalist orderly paradigm or its antithesis disorder/post-modernism, but acts like a synthesis or bridge between these two and creates a new framework for reconciling these opposing positions. Basically, both the orderly and disorderly frameworks are equally flawed.

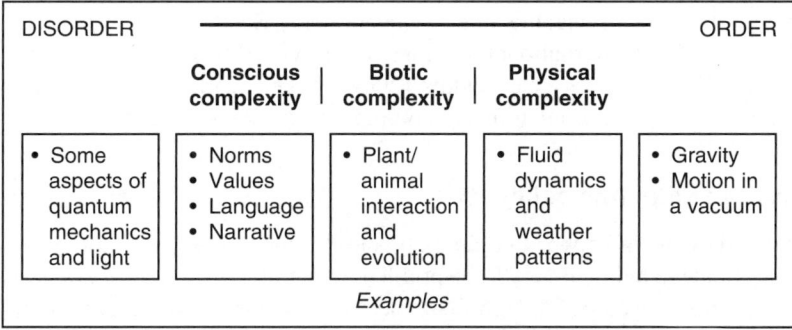

Figure 1.5 The range of physical, biotic and conscious phenomena.

Both assume that humanity and its relationship to nature are inherently orderly or disorderly when in reality they are both.

For the social sciences if one accepts a complexity framework then one must abandon the rigid divisions and certainties of both modern and post-modern science and recognise the integrative nature of the physical and social sciences. Complexity theory argues that physical and social reality is composed of a wide range of interacting orderly, complex and disorderly phenomena. One can focus on different aspects, orderly (gravity or basic aspects of existence: life/death), complex (species evolution or institutional development) or disorderly (random chance or irrationality) but that does not mean that the others do not exist. Consequently, complexity theory demands a broad and open-minded approach to epistemological positions and methodological strategies without universalising particular positions or strategies. As Richardson and Cilliers argued:

> If we allow different methods, we should allow them without granting a higher status to some of them. Thus, we need both mathematical equations and narrative descriptions. Perhaps one is more appropriate than the other under certain circumstances, but one should not be seen as more scientific than the other.
>
> (Richardson and Cilliers 2001: 12)

These conclusions, 'bridge the old divide between the two worlds (of natural and human sciences) without privileging the one above the other' (Richardson and Cilliers 2001: 11).

More practically, since the 1990s complexity has established footholds in all of the major areas of social science. To name just a few, in philosophy and social theory see: Byrne (1998), Cilliers (1998), Rescher (1998) and Urry (2003). In economics see: Barnett *et al.* (1989), Day and Samuelson (1994), Hodgson (1997), Mirowski (1994) and Ormerod (1994 and 1998). In organisational and management theory see Stacey (1999) and Stacey *et al.* (2000). In sociology and politics see: Cioffi-Revilla (1998), Eve *et al.* (1997), Elliot and Kiel (1999), Kiel and Elliott (1997), King (2000), Rycroft and Kash (1999), Smith and Jenks (2006) and Walby (2007). In development theory see: Rihani and Geyer (2001) and Rihani (2002). In political theory see: Geyer (2003), Scott (1998), Richardson and Cilliers (2007). In international relations see: Jervis (1998), Harrison (2006), Kavalski (2007). In history see: Gaddis (2002).[13] The work of these authors informs Table 1.5 on the foundations of complexity and social science.

Complexity and public policy

Given this relationship between complexity and the social sciences, how would complexity relate to public policy? The problem with this question is that we start to enter unexplored waters. Though there have been a number of recent policy-related works, particularly in the field of health, there is no general complexity theory on public policy. A major exception is the excellent work of Paul Haynes (2003) which focuses on public service processes rather than public policy as a whole. Instead, what has been emerging since the 1990s has been a blossoming of complexity based works in a variety of policy sub-areas. Examples include work on social policy by authors such as Blackman (2006), Byrne (1998); in education policy, Haggis (2007) and Tosey (2002); economic policy Beinhocker (2007) and Ormerod (1994 and 1998); in health policy Kernick (2004) and Sweeney (2006) to name just a few.[14]

All of this activity shows how quickly complexity has moved from the fringes of social science theory to the day-to-day of policy decision-making. Using these works and building on Table 1.5 we can develop a simplified foundation of a complexity-based public policy, presented in Table 1.6.

Table 1.5 The foundations of a complexity social science

Theoretical basis

- Partial order
- Predictability and uncertainty
- Emergence
- Reductionism and holism
- Probabilistic
- Interpretation

Expectations

- Over time human knowledge increases, but physical, biological and human phenomena are unpredictable and evolve into new patterns within general boundaries. *Social scientists can know more, but the systems they are observing do not stand still, are unpredictable within general boundaries and are constantly evolving and reinterpreting themselves.*
- Knowledge is powerful and useful and more knowledge can be more powerful and more useful. *However, due to the fundamentally unpredictable, probabilistic, emergent and interpretive nature of human existence, no way to know the final order. Knowledge is always limited and learning never stops.*
- Greater knowledge does not guarantee greater prediction or control. *Those with greater knowledge must constantly recognise the limits of their knowledge and must act democratically rather than in an authoritarian fashion.*
- There are general boundaries to all phenomena, but a huge and evolving range of variation and emergence within those boundaries. *No endpoint but a continual search for change within a bounded but emergent framework.*
- There is no fundamental hierarchy of knowledge or methods in the social sciences. *However, certain methods are more appropriate for some phenomena than others.*

Implications

- Researchers must taken an open-minded and flexible approach to the orderly and disorderly foundations to all phenomena.
- There are continual bounded and emerging limits to human knowledge.
- Researchers can obtain some degree of predictive and experimental results, but must often combine them with uncertainty and interpretation.
- Recognising the strengths and weaknesses of quantitative/qualitative methodological strategies and possibly combining them is the primary methodological strategy.
- Creation of an understanding of fundamental boundaries combined with
- an acceptance of continual discovery and openness is the ultimate goal.

Continued

Table 1.5 Cont'd

Vision of history and progress

- At best, social scientists can explore the interaction and continual balancing between the fundamental boundaries of our physical, biological and social worlds and the huge variety of unpredictable emergent outcomes. However, exact and detailed prediction is extremely limited (except when discussing the boundaries of the system).

Table 1.6 The foundations of a complexity public policy

Theoretical basis

- Partial order
- Predictability and uncertainty
- Emergence
- Reductionism and holism
- Probabilistic
- Interpretation

Expectations

- Over time human knowledge increases, but physical, biological and human phenomena are unpredictable and evolve into new patterns within general boundaries. *Policy actors can know more, but the systems they are observing do not stand still, are unpredictable within general boundaries and are constantly evolving and reinterpreting themselves.*
- Knowledge is powerful and useful and more knowledge can be more powerful and more useful. *However, due to the fundamentally unpredictable, probabilistic, emergent and interpretive nature of human existence, there is no way for policy actors to know the final order. Knowledge is always limited and learning never stops.*
- Greater knowledge does not guarantee greater prediction or control. *Policy actors with greater knowledge must constantly recognise the limits of their knowledge and must act democratically rather than in an authoritarian fashion.*
- There are general boundaries to all phenomena, but a huge and evolving range of variation and emergence within those boundaries. *No endpoint but a continual search for policy change within a bounded but emergent framework.*
- There is no fundamental hierarchy of knowledge or methods in the social sciences. *However, certain policy methods are more appropriate for some phenomena than others.*

Implications

- Policy actors must take an open-minded and flexible approach to the orderly and disorderly foundations of all phenomena.
- There are continual bounded and emerging limits to human knowledge and public policy despite the exponential increase in evidence/data.
- Policy actors can obtain some degree of predictive and experimental results, but must often combine them with uncertainty and interpretation, at best, probabilistic strategies.
- Recognising the strengths and weaknesses of quantitative/qualitative methodological or evidence based and interpretive strategies and balancing them against each other is the primary methodological strategy for supporting reasonable public policies.
- Creation of an understanding of fundamental boundaries combined with an acceptance of continual discovery and openness is the ultimate goal. The key isn't to find the final order and implement it, but encourage the actors in the policy area to adapt and adjust to the continual evolutionary changes.

Table 1.6 Cont'd

Vision of history and progress

- At best, policy actors can pursue a continual balancing of probabilities in a bounded evolving situation. Enabling local actors to maximise their complexity within a stable framework creates the greatest likelihood for healthy evolution and adaptation. Exact and detailed policy prescription is extremely limited (except when discussing the fundamental aspects of the policy framework).

Table 1.7 Summaries of orderly, complex and disorderly public policy perspectives

Orderly public policy	*Complexity public policy*	*Disorderly public policy*
Theoretical basis		
• Order • Reductionism • Predictability • Determinism	• Partial order • Reductionism and holism • Predictability and uncertainty • Probabilistic • Emergence • Interpretation	• Reality and rationality are relational and experienced differently depending on specific cultural and temporal dynamics • Reality is unpredictable, irreducible and indeterminate
Expectations		
• Central policy actors are able to understand more and more about their societies and humanity in general.	• Central policy actors are important, but their detailed knowledge is always limited.	• Central policy actors are irrelevant. All policies are contested.
• More knowledgeable 'evidence based' policies will create more order.	• More knowledge is useful, but may not lead to better policies or order.	• More knowledge is irrelevant. All policies can be viewed from different perspectives and many of these perspectives are irreconcilable.
• All public policy should strive to reach this point. When it is reached all fundamental policy/societal change will stop.	• There is no end to the complex nature of human development, a mix of fundamental boundaries, stability and change.	• There is no end and no direction, no way of knowing the appropriate policy.
• Duplicating traditional scientific knowledge and methods is the primary justification of orderly public policy.	• A flexible mix of traditional scientific and more qualitative and interpretive policy methods is the most effective strategy.	• Policy certainty is an indicator of rigid thinking and potential oppression.

Continued

Table 1.7 Cont'd

Orderly public policy	Complexity public policy	Disorderly public policy

Strategic implications

- Public policy actors must look for and act on rational foundations in all public policy areas.
- There are no limits to human knowledge and public policy. The only constraints are effort and technology.
- Public policy actors can obtain predictable, verifiable and repeatable policy outcomes.
- Duplicating orderly natural science methods is the 'gold standard' of public policy methods.
- The creation of an improved and stable order is the ultimate goal.

- Public policy actors must continually and flexibly combine rational and interpretive strategies in all public policies.
- Continual limits on public policy actor knowledge and actions, despite exponential increase in evidence/data.
- At best, public policy actors can obtain probabilistic policy outcomes.
- Public policy actors must continually and flexibly combine rational and interpretive strategies in all public policies.
- The key isn't to find the final order and implement it, but encourage the actors in the policy area to adapt and adjust to the continual evolutionary changes in their areas.

- Oppose modernist policy claims.
- Undermine modernist policy assumptions.
- Deconstruct modernist policies and recognise the role of the 'other'.

Vision of progress/history

- Over time, with appropriate policies and power, policy actors can move societies from disorder to order and progress them towards a final ultimate order.

- Continual balancing of probabilities in a bounded evolving situation. Enabling local actors to maximise their complexity within a stable framework creates the greatest likelihood for healthy evolution and adaptation.

- Policy merely responds to relational flows. No foundation upon which to base progress or direction to history.

Now, combining and compressing these figures gives us a convenient summary of the various fundamentals of our three perspectives, presented in Table 1.7.

What we have tried to do in this chapter is summarise the dominant perspective of twentieth-century public policy and its foundation on four rules of order, predictability, reductionism and determinism. We then went on to explore the major challenger to this position, post-modernist/disorderly perspective, and argue that neither fully captured the complex reality of everyday policy-making. From here we argued that the only way to move forward from these two positions would be build up from a complexity perspective, a new paradigmatic world view of science and society that combines both order and disorder. Hopefully, this has been interesting and is starting to make 'connections' in your mind. In order to clarify the complexity perspective and make it more practical for policy-oriented researchers and actors we now need to turn to the core concepts and tools of complexity and start exploring just what a complexity perspective can do.

2   Concepts of complexity

> To see a world in a grain of sand and a heaven in a wild flower Hold infinity in the palm of your hand And eternity in an hour.
> (William Blake, 'Auguries of Innocence', *The Pickering Manuscript* 1863)

As Blake so poetically points out, concepts are the keys to our basic understanding of the world. With this in mind, this chapter will present, very briefly, some of the key concepts used in complexity science and argue that they provide a better understanding of complex adaptive systems involving agents endowed with awareness as well as the will to act independently (i.e. people).[1] As in the previous chapters, we have chosen to explore the key aspects of complex systems by showing how they build from the physical to the biotic and then to the conscious levels. The fundamental message is that action is certainly possible in these complex arenas but appropriate styles and methods of management are required coupled with mature awareness of the limitations on predictability and control. In the following chapter, we explore how these concepts are turned into simple tools for applying complexity to the human realm.

Key concepts of physical complex systems

Lacking biotic evolutionary and conscious capabilities, physical complex systems are the simplest and most fundamental of the three types of complex systems. This means that their basic features, given below, can be found in all other complex systems:

- Limited compressibility and irreversibility
- Attractors
- Local interactions, connectivity, and simple rules
- Local variety and global stability
- Non-linearity

Limited compressibility and irreversibility

How can one differentiate between a physical orderly system and a physical complex system? The most fundamental test is to examine the regularities in the system and the length of message needed to specify their occurrence in order to describe the system (Gell-Mann 1994: 16–17). An orderly system is highly regular; a short, *compressed* description of the few recurring regularities is all one needs to give a full account of that system, hence the ease of modelling linear phenomena. Reductionist analysis; in which the behaviour of the constituent elements is used to predict the behaviour of the whole system, is appropriate in this instance. For example, ABABABABABABABABABAB can be described easily as 'AB repeats 10 times' or 'AB × 10'. It is obvious what comes next; hence the deceptive temptation to treat all phenomena as orderly systems. A totally disorderly system, at the other extreme, has very little regularity or none at all. A collection of random letters, for instance, could only be described with a similar set of letters. The system is therefore described as *incompressible*. Hence, reductionism is impossible with this system.

A complex physical system, falling somewhere between orderly and totally disorderly systems, has regularities that could be summarised, *compressed*, and other aspects that are random and, therefore, *incompressible*; a mix of order and disorder. In this case, our letter example could look something like ABTJIABOEYABMDNABPKN. There is order within the sequence that can be compressed (AB repeats regularly) and disorder (random letters between AB). Hence, this line of letters has *limited compressibility*.

Examples of physical systems could easily be found in fluid dynamics. An orderly system would be an unobstructed stream that flows at a steady rate. It is easily described and its actions are predictable. Add disturbances (rocks and sticks) to the stream and vortices begin to develop. The particles of water and the vortices themselves interact with each other to create turbulence. The flow and motion of the water becomes irregular and a complex system is created. Randomly alter the flow of the water and the vortices disappear: a disorderly system is created that would be virtually incompressible.[2]

Linked to the concept of limited compressibility is *irreversibility*. In orderly systems actions can run forwards as well as backwards and are repeatable. A constant flow of water in a pipe simply depends on the volume of water, the diameter of the pipe and the angle at which the pipe is tilted to the horizontal. The flow is not affected by the orientation of the pipe and the experiment could be repeated at will. Time is not a consideration in this system By contrast, the precise conditions of a stream in which water flows through obstacles to create turbulent regions and vortices could not be repeated. Complex systems, even in their relatively simple physical form, alter themselves over time in unique ways that cannot be reversed nor repeated. Here time is highly significant: there is a past, present and future, and conditions are different at each phase. Moreover, the past affects the present and future to a greater or lesser degree. The 'arrow of time' (Gould 1989; Coveney and Highfield 1990) is a significant element in any complex situation.

As we shall see, limited compressibility and irreversibility play a major role in conscious systems. Looking for regularities is necessary and useful but observing elements of the systems individually does not help much in determining the mode of behaviour of the whole system. Moreover, it is dangerous to draw conclusions about the behaviour of a complex system from a few instances; a practice that was popular with International Monetary Fund (IMF) mission teams who felt able to make sweeping recommendations about the political economy of a country from a fleeting visit. Similarly, irreversibility sets definite limits on what could be done in predicting and influencing the future. In this instance, recreating the past, such as a 'golden economic age' is clearly flawed. The arrow of time does point to irreversibility but it also tells us that the past casts a long shadow on the present and future. Dealing effectively with complex situations demands deeper and lengthy knowledge of what happened in the past and what is happening at present. These norms, values and conventions are sometimes called the 'culture' of the system. As we will see in a later chapter, better understanding of the 'culture' of Iraq could have helped to minimise one of the biggest policy blunders in the young twenty-first century.

Attractors[3]

An attractor describes 'the long-term behaviour of a system' (Coveney and Highfield 1995: 424). Perhaps it is better to say that an attractor gives a clue of how the system tends to behave. Readers will readily appreciate the link between this and the question of 'culture'. Attractors are easy to describe in orderly physical systems: a pendulum swinging under gravity invariably comes to rest pointing vertically downwards; presenting a *point* attractor. Its different *states* while it is in motion gradually *drain into that basin of attraction*. A frictionless pendulum swinging in a vacuum describes a *limit-cycle* attractor. In this case, the attractor looks like a segment of a circle that envelops every single *state* assumed by the pendulum on its gyrations from one side to the other. Knowledge of the attractor enables us to compute the position of the pendulum at any point in the future; perfect predictability. Point and limit-cycle attractors occur in orderly systems that are at or near equilibrium.

Physical complex systems, biotic complex systems and conscious complex systems are not so obliging. They operate within attractors that are progressively more intricate. The numerous states of an engine controlled by a governor, or a central heating system regulated by a thermostat, when monitored individually from second to second are seen to scroll through a doughnut, or *torus*, shape. Unless something goes wrong with the arrangement, the states are confined to that envelope. An acceptable level of uncertainty is designed into the system because it is not necessary to know the exact speed or temperature at any point as long as it remains within the specified limits under the influence of negative feedback provided by the regulating mechanism.

However, if the system contains processes of positive feedback, like the amplifying feedback produced in sound systems, as well as negative feedback, then one moves into complex phenomena where basins of attraction are often referred to as *strange* attractors They do not exhibit regular symmetrical shapes and due to the interaction of positive and negative feedback mechanisms, insignificant perturbations in initial conditions could radically alter the particular course followed by the numerous states assumed by the system. A well-known example, mentioned earlier, is a 'butterfly' or strange attractor developed by the renowned MIT meteorologist Edward Lorenz.[4]

The above discussion might seem esoteric for our practical purposes, but it is essential to fully understand that states of an economic or political system might appear to follow a familiar pattern for a while, nonetheless they are liable to change radically and the trigger might be an apparently trivial event. The system suddenly moves into another attractor presenting new and unexpected behaviour patterns. In human systems, the smallest of rumours or risky behaviour or a few bankers can create havoc on the most powerful of stock markets, as happened in 2008. Or, the assassination of a minor archduke can lead to the unspeakable horrors of World War I.[5]

Local interactions, connectivity and simple rules

A complex system has, normally many, internal elements that interact to shape the overall pattern presented by the system to an outside observer. Since minor events can be so important, *local interactions* in a complex system take on a much more central role. Grains of sand falling under gravity do not act differently from a single grain under gravity. However, allow these grains to interact with each other by pilling them up on a flat disc and they will begin to exhibit complex unpredictable behaviour (How high will the pile on the disc grow? Which grain will cause the next mini-avalanche?). Similarly, put ten people on a committee and anything could happen once they begin their discussions; their 'local interactions'. However, nothing will happen if none of them were to speak and for anything constructive to materialise out of their deliberations they have to learn not to all speak at the same time. In an effective committee, *connectivity* – the way members interact with each other – falls somewhere between these two extremes. A good chairperson knows how to achieve this delicate balance. The presence of a chairperson at committees illustrates the third feature needed for a complex system to acquire stable self-organisation: a *basic framework of simple rules* of interaction. In this case the rule is that the chairperson should be allowed to regulate discussion.

Computer simulations, using Boolean networks, have been used to study interactions, connectivity and simple rules in a physical complex system. By altering connectivity and the rules that govern how elements of the network interact, uneventful order, rampant disorder or organised complexity were produced and analysed (Kauffman 1993: 36). Boolean networks have been utilised in a variety of fields ranging from understanding the dynamics of ant colonies (Coveney and Highfield 1995: 247–50), to urban development and economic dynamics (see Beinhocker 2007; Stacey 2000).

The significance of interactions, connectivity, and rules is readily obvious in relation to conscious complex systems. Parallels with debilitating order under conditions of dictatorship, or centrally planned economic systems are unmistakeable. Equally, destabilising disorder resulting from lack of sensible rules of daily interactions, as witnessed during most civil war situations, are also plain to see. Self organisation in biotic and conscious complex systems does not arise by accident, but only under certain basic conditions.

Local variety and global stability

In an engine, or a central heating system, the regulatory mechanism is designed to allow the system to operate within an acceptable range rather than a precise speed or temperature. This is not a mark of inefficiency or faulty workmanship. The conditions under which the system is expected to operate fluctuate and can only be ascertained within a range of likelihood. A well-designed regulator permits the system to assume a variety of states at different points in time, allowing for *local variety*. The states are designed to fall within an acceptable range, or in complexity parlance the *attractor*, for that system. The variety of states that differ within the limits of the attractor gives the system *global stability* in an uncertain environment.

That mode of behaviour is common in torus shaped attractors found in population dynamics of coevolving species and is highly suggestive of Adam Smith's invisible hand that seeks to provide considerable stability to the market while people chaotically 'truck, barter and exchange one thing for another' locally. In this case, the local variety of the billions of daily micro-economic transactions generally stay within the stable boundaries of basic laws of supply and demand.

Minor variations between states of the system cause some changes but these are normally contained within the attractor and the outward appearance of the system is unaltered. Occasionally, these changes might shift the system into an essentially similar neighbouring attractor. Despite the frenetic internal activity, the system still seems to be unchanging. However, positive feedback could turn minor variations into major transformations, a 'tipping point' (Gladwell 2001) capable of moving the system into an altogether different attractor that displays a new global pattern. A system that appears stable might suddenly, and for no immediately obvious reason, presents a radically new pattern of behaviour. The point to stress here, with special significance to conscious complex systems, is the difficulty in linking a specific *effect* to a given *cause*. This is particularly exasperating to those used to the tidy mode of behaviour of orderly systems. The seeming durability of the communist system in Eastern Europe that lasted many decades only to collapse within a few years is a good illustration of that concept of change.[6]

Non-linearity[7]

The last of the elements of physical complex systems, *non-linearity*, is in many ways an outcome of the combined effect of some of the other elements discussed previously. When the universe was seen as a clockwork mechanism, calculus evolved to explain the workings of the clock.[8] Any force or path could be charted with simple calculations. The quation $y = ax + b$ charts the path of any straight line on a two-dimensional graph. Plot out the different force vectors and in a clockwork universe everything, from the motion of planets to the falling of apples, could be explained. For many, mathematics was (and remains) the ultimate orderly subject. However, not all of the world or mathematics is linear. In 1845 P.F. Verhulst, a scientist interested in the mathematics of population growth, devised a mathematical equation that produced feedback on itself and non-linear properties. This equation, $x_{n+1} = bx_n(1 - x_n)$ did not produce a straight or curved line on a two-dimensional graph (see Briggs and Peat 1990: 57). Instead, it produced a series of *bifurcations* (points at which the system divides between two different attractor states) that eventually led to highly non-linear outcomes. In essence, there was not a single answer to the equation, but a whole range of answers.

This may seem detached from the daily reality of policy life. Nevertheless, the reason for including non-linearity here is simply to highlight the gross error made in treating complex phenomena as orderly systems, for which assumptions of linearity are perfectly appropriate. As discussed earlier, science is no longer just linear and orderly. Non-linear mathematics and other mathematical concepts like fractals (Lesmoir-Gordon *et al.* 2000) demonstrate that even in the most basic scientific fields complexity is having a growing influence. Consequently, when policy actors assume that the most valid and 'scientific' strategies are based on orderly foundations of causality, reductionism, predictability and determinism they are basing their views on a 200-year-old version of science and ignoring a whole range of recent developments. Hence, when dealing with complex policies or socio-economic situations, top-down management, command-and-control from the centre, and 'hard systems management' styles founded on resolute direction by a few 'inspired leaders' who define end-state plans and pursue them to a successful conclusion are revealed as fundamentally misconceived, ineffective and often counterproductive.

Key concepts of biotic complex systems

Complexity theory and thinking has a long and deepening history in the biological sciences. Many of its earliest thinkers and most ardent supporters are found in these areas.[9] For many in the biological sciences, complexity is simply 'commonsensical'. The evolving and adaptive capabilities of biotic complex systems provide a new level of complexity to add to that discussed above in relation to physical complex systems. The most important for the present purpose include:

- Adaptation, survival, variety and 'good enough'
- Evolution

- Punctuated equilibrium, gateway events, and frozen accidents
- Arrow of time and depth.

Adaptation, survival, variety and 'good enough'

As we saw in the last section, Boolean networks have been used to study the cyclical process of *adaptation* and *survival* in an uncertain environment. Success in adaptation depends on three obvious requirements. First, the system had to assess its environment: learning. Second, it has to act on that knowledge: response. And thirdly, and crucially, it has to survive long enough to repeat the cycle over and over again. Essentially, survival of the fittest is really survival of the most adaptive and stable (Dawkins 2006).

Observing adaptation in the living world, however, presents a number of logistical problems, the most obvious being the elapsed time between cycles of reproduction.[10] Computer programmes, like the pioneering work done by Thomas Ray, simulated Darwinian evolution by allowing replicating digital 'organisms' to compete for limited computer memory space.[11] A most illuminating outcome was the realisation that inside the computer, as in nature, and in accordance with Darwin's views, selection pressures for adaptation came from activities by other co-evolving entities. The physical environment, in this case the central processor, was the junior partner in that struggle. The similarity with interdependence and competition between states in the human world is fairly obvious.

Computer simulations mirrored activities in real-life situations, from aggression and parasitic behaviour to co-operation and competition. In every case *variety* emerged as an essential asset for survival. However, a key feature was laid bare through these studies: evolution does not lead to an optimal end-state. *'Good enough'* seems to be the sum total of what nature hopes to achieve, successful evolution being an ongoing open-ended process of often small, effective adaptations by which the entity manages to improve its performance in a changing environment (Dawkins 2006; Kauffman 1993; Rihani 2002). This implies that there is no such thing as the one superior creature or evolutionary form, merely continually evolving ones that are more or less successful in a given environmental context and time. The rise and fall of the dinosaurs, unrivalled masters of the Earth for millions of years, provides the most obvious example.

This picture of adaptation and survival coming from biotic complex systems assumes cardinal importance when complex adaptive systems involving conscious human beings as the internal interacting elements are addressed. End-state planning and the associated search for an 'end' to the progress of human society are exposed as utter nonsense. As will be discussed, life is infinitely more demanding, and exciting. There are no shortcuts, but simply a few signposts that might suggest a possible direction of travel that could avoid some of the pitfalls on the way. But crucially, there are no guarantees.

Evolution

For complexity science *evolution* is an arduous task undertaken against considerable odds. A biological system has to be able to adapt, by having at all times some elements that are fit for the prevailing circumstances, and then it has to remain stable for long enough to adapt again. This cyclic process, survival and adaptation, entails a battle against the fundamental physical laws of nature as well as a continual search for new co-operative and conflictual strategies for survival that are 'good enough'.

There are several aspects of the last sentence that seem odd on first inspection. First, how could evolution be a battle *against* the laws of nature? Isn't evolution a law of nature? Yes, for biotic complex systems it is a law. However, that law or regularity is less fundamental than the second law of thermodynamics which dictates that things, when left to their own devices, must move towards disorder and decay. Put simply, the second law is based on an unshakeable principle that disorder is a more probable state than order. This is commonsensical enough; for example, a pile of rubble never assembles itself into a building but the reverse, by contrast, is a self-evident occurrence. In short, continual energy and action are required to maintain a biotic complex system in a stable, less probable, state of survival and, hopefully, evolution. To keep it there is no mean task: too little effort and the second law triumphs by pushing it into disorder, too much effort and the system is 'killed' as unvarying order prevails everywhere; all regions 'freeze in fixed states of activity' (Kauffman 1993: 174). Living systems mastered the trick of remaining within the narrow layer of complexity, between order and disorder, after millions of years of trial and error.

The next aspect of that earlier statement is the 'co-operative and conflictual' nature of evolution. How can it be both? Isn't evolution a fight to the death that is, as Tennyson wrote 'red in tooth and claw'? For biotic systems, a whole range of co-evolutionary adaptive strategies exist to pursue survival and reproduction. Conflict is only one of those strategies. It is important to note that one of the most fundamental co-operative strategies for survival is sex. Sexual reproduction introduces a key mechanism for change in chromosomes other than random mutation.[12] Other co-operative forms; for example, organised colonies of insects, symbiotic behaviour in plants and animals, are too numerous to mention.

Lastly, 'good enough' is particularly important to the above sentence for clarifying some of the distortions of a belief in evolutionary order. When Darwin freed evolution from the hands of God, the Newtonian framework made Nature into a clockwork mechanism and demanded that it should be linear and progressive. From slime mould, to animals, to humans evolution led to the creation of the 'fittest'. Rather than seeing the growth of biological diversity as a beautiful treelike flowering of complex creative power, it became a hierarchical pyramid with human beings at the top. Even worse, this orderly evolutionary logic was applied to different races, ethnic groups and gender resulting in some of the most vicious and brutal treatment in the history of humanity.[13] From a complexity perspective, although evolution 'produces organisation and interactions that are highly refined, they are invariably still improvable and not truly optimal' (Coveney and Highfield 1995: 119). As we will argue in later chapters, evolution is not about finding the final order or 'way', but a continual search for and evolution towards the next broad range of 'good enough' ways.

Punctuated equilibrium, frozen accidents, regularities and gateway events[14]

The adaptation and evolution of biotic complex systems is an uneven process that does not follow a smooth line of tidy progress from one stage to the next. Generally, evolving biotic systems stay more or less as they are for relatively long periods of time and then undergo fast radical change. This pattern of large upheavals separated by long periods of global stability, but energetic local activity, is referred to as *punctuated equilibrium*.[15] The interval of apparent inactivity is deceptive. Ripples of change come and go without leaving a discernible trace, an individual or group dies or survives, but occasionally relatively minor events ('butterfly effects') succeed in shunting the system into another attractor. Subsequently, in a stable system life quickly settles down to the new pattern. Variety, and hence flexibility, allows the system to bend with the wind (the emergence of a new predator, competitor, etc.) and avoid the only truly orderly state in the evolutionary process, extinction (Rihani 2002).

Punctuated equilibrium is easily observed in the affairs of humankind, as the rise and fall of ancient empires and hegemonic powers clearly shows. In modern history, there have already been three lengthy periods of stable hegemony: the United Provinces, mid-seventeenth century; the United Kingdom, mid-nineteenth century; and the United States of America, mid-twentieth century (Wallerstein 1983). The abrupt fragmentation of the USSR after three-quarters of a century of stability is of course the ultimate recent illustration of punctuated equilibrium in action.

In natural evolution, chance events from the past sometimes become an integral part of life from that point on. For obvious reasons they are referred to as *frozen accidents* (Gell-Mann 1994: 133). The formation of carbon-based form of life on Earth is a good example. Once that frozen accident was built into the system, the future of all creatures on Earth was set on a distinctive path. Other fundamental processes included the leap from single- to multi-cellular creatures, the creation of sexual reproduction and later the formation of self-awareness. As frozen accidents accumulate they create more general *regularities*. Following the Big Bang, multiple frozen accidents led to the regularities that created the Milky Way. With the success of mammalian life, multiple frozen accidents led to the regularities of Neanderthals and *Homo sapiens*.

In our human world, this combination of frozen accidents and regularities can easily be seen in the current structure of many of our major institutions. The European Union was a product of a wide range of frozen accidents (chance meetings between core diplomats, unpredictable aspects of policy spillover, etc.) and entrenched regularities (the Cold War, US economic hegemony, etc.). The unending contest between the conservatory influences of frozen accidents and emerging regularities, on the one hand, and adaptation, on the other, typifies all biotic and conscious complex systems.

Similar to frozen accidents, *gateway events* are a major factor in creating the pattern of punctuated equilibrium. Unplanned gateway events crop up from time to time to open up niches that promise new and unexpected opportunities for some and disaster for others. That happened at the end of the Permian period, 245 million years ago, when more than half of all species on Earth disappeared, and during the Cretaceous extinction, about 185 million years later, when dinosaurs became extinct along with one third of the world's animal and plant life. The rest of creation, organisms able to tolerate or benefit from the new conditions, mammals in particular, flourished in abundance (Rihani 2002).

Gateway events in natural evolution have their social, economic and political counterparts. The Industrial Revolution and, later, the invention of the internal combustion engine and the Internet are perfect specimens of gateway events, comprising the emergence of new developments and the subsequent steady filling of the niches they bring forth. But a system, an individual, group or nation must have sufficient variety and flexibility to survive the shock of the latest gateway event in the first instance before it could proceed to profit from the new opportunities on offer.

Arrow of time and depth

As already mentioned in the context of physical complex systems, the flow of time is highly significant when it comes to all complex phenomena including biotic and conscious systems. The *arrow of time* moves in one direction from past through present to the future.

Progression from past to future imposes certain constraints. As described above the system goes through many minor and some major changes through its history. Frozen accidents accumulate as regularities that define a direction of travel which itself might be influenced by gateway events. The pattern of a system at any instant is the product of its eventful past up to that point. Future course of events depends on many factors, which judging by what happened in the past imposes strict limits on prediction. Moreover, as conceived by Stephen Gould, in biotic complex systems one cannot replay the tape of time because 'any replay of the tape would lead evolution down a pathway radically different from the road actually taken' (Gould 1989: 51).

This arrow of time implies that biotic and conscious systems evolve towards increasing complexity and it takes time for that build up to occur. That imposes yet another constraint: many layers of interconnections and adaptations have to come into being at a pace that could not be accelerated appreciably. As we will argue later, money, imported expertise and grim determination to change a social system (neighbourhood, city or country) are no substitutes for the time needed to generate *depth*.

The above depiction of a turbulent and uncertain timeline suggests that for biotic complex systems (and for conscious systems as well) complexity has been gaining *depth* for many millions of years. In short, 'life is becoming more complex' all the time. This often-heard lament is based on scientific fact rather than pure nostalgia. Complex systems, those that adapt and survive at least, acquire more and more complexity; and the more complex have a tendency to gain more complexity than other less complex systems.[16] It is important to stress that the development of greater complexity is only a tendency. In specific instances 'complexity can either increase or decrease', but 'the greatest complexity represented has a tendency to grow larger with time' (Gell-Mann 1994: 244). Hence, while the lengthy process of evolution unfolds there is a high probability that the average complexity of all biotic systems will also increase.

In conscious complex systems, there is a similar tendency. As human history marches forward, frozen accidents (the usefulness of trade for example) create new regularities (the market) that produce gateway events (the Industrial Revolution) which add depth to the totality of experiences of conscious complex systems (human history). As in biological systems, there is no guarantee that those systems with the most depth will be the most likely to survive and succeed. During the Cold War, indigenous tribal groups in isolated regions with fundamental survival skills would probably have been much more likely to survive a nuclear holocaust than the average New Yorker or Moscovite. Nevertheless, for the vast majority of situations the more diverse and adaptive city dwellers with much greater complexity depth are much more likely to survive and prosper. The emergence of hegemonic powers among nations and the inclination for the rich to get richer on a global and national scale hint at social, political and economic similarities (Kennedy 1989).

Key concepts of conscious complex systems

So far we have stepped through the concepts of physical and biotic complex systems. That progression to our main focus, conscious complex systems, was essential in outlining the main features that characterise systems in the social, political and economic fields. This is not a new departure: for a growing number of researchers, complexity not only helps to explain the biological world, but can also act as a map or model for understanding how the brain works as well as the formation of consciousness.[17] In this book, however, we concentrate on the practical consequences of dealing with human activity and complex systems and the implications of that approach on policy-making and actions. Mainzer summed up the key difference between that and the traditional approach of assuming all systems to be orderly:

> The complex system approach discloses that emergent effects of the whole system are system effects which cannot be reduced to the single elements. Philosophically, the whole is more than the sum of the parts
> (Mainzer 1997: 124)

Human beings have attributes; such as the ability to store, process and communicate knowledge and the capacity to build narratives and scenarios to test hypotheses. Their activities, therefore, lead to the emergence of complex interactions with greater depth that go well beyond those seen in physical and biotic complex systems. Some of the most important concepts that arise from the distinctive capacities of conscious complex systems include:

- Bounded freedom and diversity
- An evolving societal framework
- Emergence and unpredictability
- The limits of knowledge and importance of learning.

Bounded freedom and diversity

As conscious beings with advanced capacities for information, communication and interpretation we are in a position to know and control ourselves and surroundings to a much greater degree than other biotic phenomena. Unsurprisingly, this gives us a significant degree of control and puts humans in a potential state of enhanced *freedom*, relative to our biotic competitors. This freedom is both a blessing and a curse. As Jean Paul Sartre wrote in *Being and Nothingness*, 'I am condemned to be free'. Freedom gives us the ability to go beyond the limitations of genetic compulsion. Survival is important, but love and hate, pride and joy, drive us as well. At an extreme, given more than just survival instincts humans may choose individually or collectively to destroy themselves. The examples are multitudinous from individual suicide (a remarkably common human trait), to group death (the mass suicide of an American religious cult in Jonestown, Guyana in the 1980s) to a willingness to accept almost complete species suicide (nuclear holocaust). There is no law that guarantees that humans would not choose the ultimate orderly attractor state: death or extinction. This freedom gives humans the ability to perform horrific, sometimes seemingly illogical, acts and pursue wonderful dreams and lies at the base of our remarkable diversity and adaptability.

Diversity and adaptability are nothing new to biotic phenomena. Multitudes of species do it every day without any advanced thought or consciousness. Genetic mutation, natural selection, competition, reproduction and survival all enforce diversity and adaptation on living creature and the complex systems they inhabit. What makes human diversity and adaptability so different is that their significant abilities to store, communicate and interpret information enable them to radically affect their diversity and adaptability. Genetically, humans are relatively homogeneous creatures, particularly when compared to the variety of monkeys or for that matter, insects or simpler forms of life. Physiologically we are remarkably similar and yet our enhanced conscious abilities allow us to survive in almost any natural environment and to dominate if not control the other creatures on Earth. Moreover, these abilities give us the capacity to diversify our actions and interpretations of ourselves and environment in a multitude of ways. Not only do our greater storage, communicative and interpretive capacities increase our potential diversity of actions beyond the boundaries of our natural competitors, but they significantly increase the range and speed of our adaptation to new threats and opportunities. Our ability to adapt, respond to and exploit our environment is so advanced that our future is less determined by our biological competitors than it is by the actions of our own species.

Nevertheless, as seen in physical and biotic complex systems, too much diversity can lead to negative outcomes as well. If each individual pursues completely autonomous actions, advanced human relations would quickly break down. How could one communicate if one could not agree with others on the fundamentals of language? Likewise, if common human experiences were not interpreted in a similar fashion, how could one create a collective understanding? How would others learn from and adapt to that shared experience? Here, if you will, lies the relativist nightmare. Human capacities enable us to obtain an enormous amount of diverse actions and interpretations. However, without some level of commonality or *boundaries*, complex human interaction experiences a complexity catastrophe and the gains of human consciousness are lost. Without some form of *evolving societal framework*, human actions and interpretations begin to resemble random noise and do not allow for the creation of emergent stable patterns.

Evolving societal framework and the role of elites

What do we mean by an evolving societal framework? Since humans have evolved the capacity for self-awareness, they have increasingly freed themselves from some of the constraints of nature. For example, with the creation of agriculture in the Fertile Crescent, the strip of land between Egypt and Turkey, around 10,000 years ago humans freed themselves from the limits of a nomadic hunter-gatherer existence. Elaborate social hierarchies and mythologies emerged and led to detailed reproductive and sexual norms and habits (Morowitz 2002). Later, the movement of goods and trade enabled cities to emerge at greater and greater distances from food sources. Civilisations and empires quickly developed and the modern world was born. The details, of course, varied dramatically. However, most fundamentally, one can think of this evolving societal framework as the accumulation of frozen accidents and regularities in human development that has a tendency to become more complex with time.

The growth of conscious complexity and an evolving societal framework came with three costs. First, with this increasing freedom from biological competitors and constraints came the growing reliance on social norms, beliefs and customs to maintain the complex human systems. As human societies, like other complex systems, developed more regularities they became increasingly complex. This growing complexity made individuals in the society progressively more interdependent. Economic activities and skills became increasingly specialised, social norms became more detailed and individual humans became more and more dependent on human society. Soon their very lives depended on the general acceptance and success of the social framework of their given society. For a growing number, the collapse of society meant individual extinction.

Second, as civilizations and transportation developed, the fate of any evolving society became increasingly dependent on interaction with other societies. In some cases, this interaction brought tremendous benefits as demonstrated by the success and popularity of the ancient Indo-European Spice Routes or the intellectual gains brought to Europe by the Arabs, sparking the European Renaissance. More recently, the benefits of trade between the advanced industrial countries in the past few centuries have been incalculable. In other cases, this interaction led to destructive competition and aggression. For example, the various indigenous tribes of North America and advanced civilisations of Central and South America, particularly the Aztecs and Mayans, had developed successful societies that had survived for thousands of years in hostile natural environments. However, when confronted with the more technologically advanced civilisations of Western Europe, they were easily overcome and in some cases completely wiped out. Innumerable local and regional wars since that time, two world wars in the twentieth century and the threat of nuclear holocaust clearly demonstrate the negative aspects of human interaction.

Third, human societies, despite their tendency towards greater complexity, have demonstrated a tendency towards a stultifying elite rigidity. This is influenced by an innumerable number of contingent factors, but grounded in the self-interest and *role of elites*. Fundamentally, as societies become more complex and social groups more stratified, elite groups tend to develop strategies and narratives to maintain their dominant positions.[18] In pre-Enlightenment times, these arguments were generally based on military prowess and/or the elites' relationship to some form of higher being. In more recent times, the basis of elites rested on a varying form of 'social contract'. Thomas Hobbes (1588–1679) argued that his idea of the 'social contract' and the need for a 'sovereign' to impose order amongst individuals was an inherent part of human existence in an overarching brutal state of nature. Various models of democratic social contracts such as republican, parliamentary, federalist, and so on (to name just a few Western examples), have all evolved to try and balance the need for an elite to help maintain the larger societal framework and to counter the tendencies of elites to lock themselves into place and become overly rigid.

What this implies is that, despite elite protestations to the contrary, there is always a tension in human complex systems between rigidifying elites and normal social complexity and chaotic interactions. Obvious large-scale political examples of elite rigidty include the actions of various maniacal dictators that infect much of the developing world. Robert Mugabe's 2008 farcical election in Zimbabwe is only the most recent in a string of these types of examples. At an economic level, similar processes can be found in the relationships between monopolistic global corporations exercising massive power and control over poorer regions and populations. As we will see later, how one balances this tension is a key factor in societal success and failure.

Emergence and unpredictability

Emergence has become a major concept linked to complexity and the title of an excellent journal for complexity and the social sciences.[19] As summarised wonderfully by Peter Coveney and Roger Highfield:

> Life is also an emergent property, one that arises when physicochemical systems are organized and interact in certain ways. Similarly, a human being is an emergent property of huge numbers of cells, a company is more than the sum of its pens, papers, real estate, and personnel, while a city is an emergent property of thousands or millions of human beings.
>
> (Coveney and Highfield 1995: 330–31)

Complexity and emergence make the possible directions of the evolution of humans and human activity systems wonderfully open. This openness, non-linearity and constant adaptation contribute to the *unpredictability* of these systems.[20] This does not mean that the conscious world is totally disorderly. One can predict certain fundamentals with a high degree of confidence in the long term. The world will most likely continue to be divided among nation-states in the next 50 years. Economic leadership will fluctuate significantly during this period. One can predict certain details in the short term as well. If it is a Sunday, most workers in a Christian world will go to work the following day. Beyond these generalities, however, the picture for conscious systems is decidedly fuzzy.

What is bizarre and dangerous is the continued belief in assumptions of and methods for dependable long- and short-term predictability. This general tendency has two particular manifestations: the unflinching belief in the 'core model' and ability to copy the new model. As extensively discussed in Rihani (2002) and others, for much of the post-WWII period there has been the belief in a core model of economic relations based on a so-called 'Washington consensus'. This core model, based on free markets, international trade and minimal state intervention was the bedrock of mainstream economic and developmental thinking for over 50 years. And yet, as demonstrated by the Cambridge economist Ha-Joon Chang (2003 and 2007), all of the major 'advanced' countries never followed the core model, but cheated on it (protectionism, tariffs, etc.) and adapted to it (state support for industries, technological research incentives, etc.) in a multitude of ways. In the end, the ones who suffered were the ones who rigidly tried to (or were forced to) follow the core model.

Similarly, there is the continued belief that a new 'final' model can be found and copied. In the 1980s, the champions were the Japanese and South Koreans who seemed to be the world beaters. 'Just in time' management and stable lifetime employment were the keys to economic success. In the 1990s, the booming USA economy was again the idol stressing its openness and flexibility. In the 2000s, the rise of China and India has reshaped the global debates. In all of these cases, experts at the time were certain that if their countries or systems could just copy the ways of these wonderful cases and import them exactly (like a new gear in a clockwork mechanism) they would be able to duplicate their success.

The obvious problem with this from a complexity perspective is that each complex system has its own emergent properties and unpredictability. Rigidly adhering to the precepts of the 'core model' or to the fluctuations of the current 'leader' are equally poor strategies, unlikely to result in stable developmental outcomes. As we will argue later, this applies not just to economic strategies, but to all other policy areas at every level (international, national and sub-national).

The limits of knowledge and importance of learning

The above inherent indeterminacy sets obvious limits on what is possible to predict when dealing with conscious complex systems. Undoubtedly, this constraint is most uncomfortable when viewed in today's dominant orderly frame of reference. However, it is also unavoidable. The certainty associated with reductionist scientific methods is largely spurious when applied to these complex situations. The danger lies in trading real but admittedly limited predictability for chimerical certainty. Explanation and prediction in the case of such systems centres on their stable global properties rather than the details of what is going on inside. That gives an important clue as to the appropriate actions that could be taken to achieve positive results. As we will argue in more detail later, command-and-control methods are useless in complex situations. They might succeed temporarily when applied with sufficient force but they are not sustainable as long-term policies. Conscious complex systems respond better to light-touch styles of management based on constant monitoring of overall patterns of performance coupled with judicious small-scale incremental adjustments.[21]

Does it matter that the 'experts' persist in their efforts to predict the unpredictable? Sadly, yes it does. At best, their efforts might have a neutral effect on events. To an extent, their pronouncements serve a purpose of sorts as they supply the reassurance sought by some elites and non-elites. In previous eras priests and soothsayers performed that task, but nowadays people demand rational answers, and more importantly quick fixes, based on some sort of science no matter how imperfect. Regrettably, when words are translated into actions the results often prove tremendously harmful from misguided national and international state policies to unreasonable social expectations and destructive group conflicts.

So, are we at *the limits of knowledge* as some authors argue (Horgan 1996)? Have we discovered all there is to know and what is left is primarily beyond our reach? This argument rests heavily on a reductionist framework, that there is a given set of fundamental laws and that once we have found them the story stops. For complexity there is no end. Basically, all of the potential for human conscious complexity was contained within the singularity of the Big Bang 12 billion years ago, but knowing the fundamental laws of physics would never have given one the ability to predict that the sport of basketball would emerge in a particular culture in the late nineteenth century on the third planet from a star in the Milky Way galaxy.

Similarly, if we are beginning to see the first glimmers of a new type of physical consciousness in artificial intelligent machines and supercomputers how could we possibly predict the long-term evolution of that emergence.

Lastly, complexity implies *the importance of learning* as a process rather than an endpoint. As we will argue in detail later, modern education was built on a reductionist framework. Knowledge could be partitioned with clear endpoints, we call qualifications, certificates or diplomas. In this sense, learning is separate from being or doing. From a complexity perspective, learning is a continual life-long process of individuals and systems. Learning how to learn becomes much more important than the memorisation of 'facts' and 'laws'. In this sense, learning, like adaptation and diversity, is a fundamental element of conscious systems. The advanced ability of humans to learn radically lifts human potential beyond the boundaries of other biotic phenomena. As we hope to show, recognising the fundamental and continual complexity of our physical, biological and human worlds implies a perpetual openness and hence continual learning.

3 Tools of complexity

> Man is a tool-using animal ... Without tools he is nothing, with tools he is all.
> (Thomas Carlyle, *Sartor Resartus* 1834)

When we first began working with complexity and integrating it into our teaching and presentations many of our students and the policy actors who attended our talks were fascinated by the 'big' ideas and engaging concepts, but continually queried us on how can this be used in daily life and what can it do for those engaged in policy activities. This chapter is all about answering these questions. In it we have selected six key 'tools' of complexity that we find particularly useful for applying complexity ideas and concepts to everyday policies and daily life.[1] These tools will be interwoven into the subsequent policy chapters:

- Cascade of complexity
- Balance and range of outcomes
- Complexity mapping
- Fitness landscaping
- Stacey diagram
- Stakeholder involvement and soft systems methodology

Cascade of complexity

From a complexity perspective, human beings are neither a cog in a massive universal machine, nor the pinnacle of universal development. Instead, an individual human being is a complex system evolving within a larger social complex system, which is evolving within a larger biotic system, within a larger physical one. Humans must live within the boundaries of these nested systems. We can't ignore the laws of gravity or species evolution, but we have a huge degree of freedom within the boundaries of these systems. Moreover, our march through time within those boundaries is continually shaped by the frozen accidents, gateway events and regularities that we discussed in the preceding chapter.

Visualising this process is something we call the *cascade of complexity*. It is like a series of fountains stacked on top of each other. The base of each fountain

54 *Tools of complexity*

represents a fundamental gateway event that opens the system up to a whole new range of possibilities. Frozen accidents occur that shape the path of the system through the range of possible pathways. Meanwhile, the boundaries of the fountain represent the regularities that emerge that put limits on the possible outcomes. Over time, a new gateway event occurs and a new variety of pathways emerge. The uneven flow of the cascade captures the erratic nature of its punctuated equilibrium. This is not the smooth and stately flow of an orderly process leading to a clear and final endpoint. Figure 3.1 attempts to visualise this process at its most fundamental level.

As it shows, following the arrow of time and beginning with the Big Bang some 12 billion years ago, fundamental physical laws were created which set the boundaries of existing physical phenomena. These boundaries are extremely wide, but they are still boundaries. Basic physical laws, gravity for example, limit the range of physical phenomena. Presumably, as argued cogently by

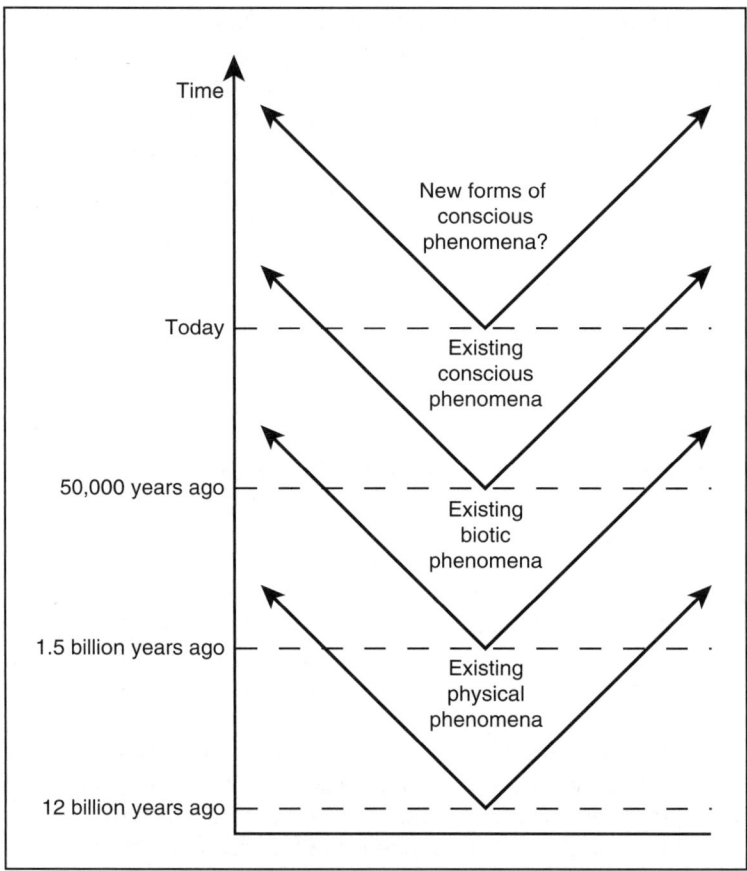

Figure 3.1 Complexity phenomena and the arrow of time.

Tools of complexity 55

Stephen Gould (1989) if the tape of the universe could be rewound and run again other physical laws and regularities might have emerged which would have reshaped the boundaries of the physical universe.

Likewise, the beginnings of life around 2 billion years ago set the boundaries for the evolution of life on Earth. As we are well aware these boundaries are gloriously broad. However, they do exist. Life as we know it requires certain chemical and physical basics. This does not mean that if the 'tape' of life was rewound or if life emerged on a different planet that it would evolve in a similar fashion. Similarly, with the emergence of hominids about 1.8 million years ago and consciousness, represented by the creation of language in the last 100,000–150,000 years (Morowitz 2002: 156), the boundaries of existing conscious phenomena were set. Again, outside of the existing set of known phenomena a range of other conscious phenomena were possible. For example, if the Neanderthals had survived what form of consciousness and conscious phenomena would they have evolved? Pushing one's imagination further, what kind of imagination would a dolphin have evolved if it had been first on the evolutionary ladder of conscious complexity? Lastly, coming to the present period, we have left open the possibility of new types of consciousness. With the growth of computer power and artificial intelligence who knows what new type of mixed physical/biotic consciousness may emerge.[2] The possibilities are both frightening and staggering. As we will see in subsequent chapters, a cascade of complexity can easily be applied to national, institutional and even individual systems and phenomena.

Balance and the range of outcomes

If we return to the orderly perspective there are two underlying possible outcomes for most phenomena: order and disorder. Intuitively from the Newtonian perspective, orderly systems and phenomena are inherently better than disorderly ones. As we saw in Chapter 1 in the words of Condorcet, the search for the fundamental laws of human society and greater human order was what the social sciences and public policy were all about. The pre-eminent focus of research and policy efforts should be to eliminate disorder and uncertainty. A recognition of some form of balance between order and disorder is akin an admission of intellectual and policy failure.

On the other hand, from a complexity perspective, humans must constantly seek the appropriate *balance* between order and disorder in their symbiotically evolving environment. Acting on and in an evolving complex system that changes with their own actions, individuals' and groups' choices become more uncertain the more detailed they become. At an individual level, we know that we must eat to survive, but what kind of food is the best for us and will we enjoy it? For a particular hospital, it must certainly have a number of key wards and specialisms, but should it focus on helping the elderly or young, cancer or diabetic patients, public health initiatives or accident and emergency care? At a national policy level, maintaining a stable interest rate is a generally good economic policy, but should it go up or

56 *Tools of complexity*

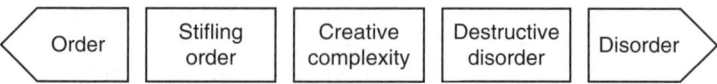

Figure 3.2 Range of outcomes.

down by a quarter of a per cent today or tomorrow? At the international level, will aid from the European Union to Russia strengthen Russia's struggling market economy or be filtered through corrupt politicians into the hands of the Russian mafia? In essence, these systems and the actors within them must continually strive for the zone of creative complexity on a broad range of potential outcomes. In a very simplistic fashion, this range can be represented in Figure 3.2.

Of the various areas or zones, the most difficult to survive in are the *order* and *disorder* zones. In the order zone, actors are stripped of their ability to make local decisions, freely interact and respond to emergent developments in new ways. This is the zone of the most rigid authoritarian and totalitarian regimes and extreme policy areas. Likewise, the zone of disorder is one where human actors struggle to make any local connections or interactions because they are so unpredictable. Periods of extreme social disorder or civil war, where all trust evaporates in a cloud of fear, is as close as humans systems can get to this zone.

Much more common are the zones of *stifling order* and *destructive disorder*. Under stifling order, for example, central actors apply overly rigid criteria on local actors who are forced to fulfil these criteria to the detriment of obvious local problems and opportunities. This can occur at all levels of human systems such as where overbearing parents force their children down a particular life pathway, education policies rigidly stipulate all daily teaching activities, health policies set multiple and rigid targets, interest rate policies are set to strict criteria, a dictator decides that a particular ethnic, social or gender group can only perform certain social or economic tasks, and so on. The same applies to destructive disorder. In this case, the system lacks a clear framework of sensible rules or suffers under constantly shifting rules. Children from broken homes going through various social and foster care systems, peasants toiling to farm in areas with competing warlords fighting for control, and policy actors struggling to make sense of fluctuating governmental demands are all caught within this zone.

The zone of *creative complexity* is generally the most productive for human activity systems. It combines a stable evolving framework that establishes core boundaries and enables as wide a variety of local interactions as possible. Systems in this zone have a general direction, but are not rigidly locked into a particular pathway. They adapt, shift and respond to evolving conditions and are allowed to explore, make mistakes and learn as they go along. Most successful individuals, policies and nations manage to find their way into this pragmatic zone and *balance* within it for lengthy periods of time.

In general, this is all blatantly obvious and simplistic. Most 'coalface' policy actors understand it immediately. The shocking aspect, as we will see in later

Tools of complexity 57

chapters, is just how often these simple points are ignored by academics and policy actors.

Complexity mapping

Building on our tool for balance and range, we can return to our box figures in Chapter 1 to develop a way of mapping the various orderly to disorderly elements of virtually any complex system. We call this *complexity mapping* and by using Figure 1.1 in Chapter 1 as a template we can produce an overview of the range of complexity dynamics of human phenomena. The key point to recognise is that there are both orderly and disorderly dynamics and that they are not hierarchically organised. A given human outcome, a decision to have coffee at breakfast or bomb a particular village, could be based on orderly, complex and disorderly dynamics with all being equally essential to the final outcome.

Beginning with order, the most fundamental and universalistic elements of human complexity are basic physiological functioning, in particular life and death. These physical boundaries and requirements, carbon-based life forms requiring air, water and food to survive and reproduce, are the most orderly aspects of human existence. Deprived of these fundamentals, a human will die. What could be more orderly?

Moving into the range of complex systems, examples of mechanistic complexity in human systems would involve situations where individuals were forced to act in a mechanistic fashion. Traffic dynamics, choosing one road or another; crowd dynamics, choosing one exit or another; electoral outcomes, choosing one candidate or another, are all examples that mimic the dynamics of mechanical complex systems. Like mechanical complex systems, relatively simple and stable patterns will emerge. However, this is no guarantee that these patterns will be continuously stable (traffic jams, crowd delays, landslide elections) nor is it possible to perfectly recreate the exact conditions of these events at a later time. The boundaries of physical complex systems apply.

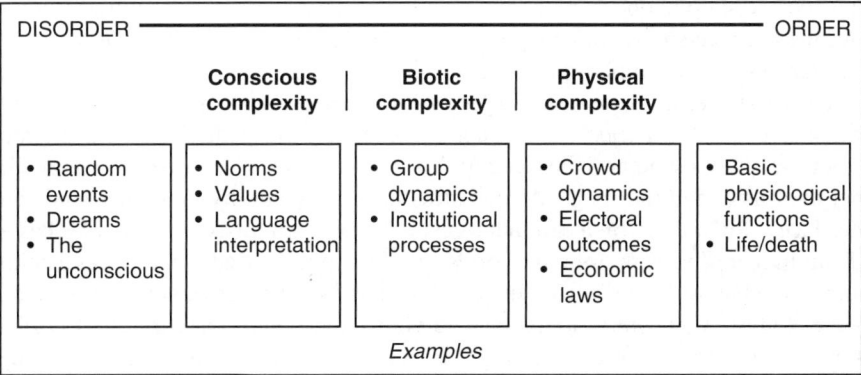

Figure 3.3 Complexity mapping for human phenomena.

Examples of processes similar to biotic organic complex systems in the human world can easily been seen in the organisational dynamics of economic and social institutions. As demonstrated by the huge growth in management and complexity literature, a business is a complex system that interacts with a larger complex environment (the market) that is very similar to the earlier model of a fish in a pond. General patterns emerge and the business is able to adapt to changes in its environment, but exact predictions and explanations of how a change in the environment will affect the business or the best strategies for the business to survive in the altered environment are extremely difficult to know in advance.

An added layer of complexity in the human condition is the faculty of consciousness. Human beings create signs, symbols, myths, narratives and discourse in order to understand, control and exchange information about their surroundings. This ability adds another layer of complexity onto the human condition that is distinctive from the natural world. Examples of this conscious complexity include the creation of language, norms and values, and discourse. An example can be taken from virtually any type of human verbal interaction. A seemingly simple student–teacher relationship can be layered in historically, culturally and personally specific aspects that would be impossible to recreate in a different time and place.

Lastly, the area of disorder plays a huge role in the functioning of human systems. Common human experiences that readily come to mind to describe this area would be the chaotic nature of dreams and the unconscious, random effects of certain disorders on the complex functioning of the brain and the phenomena of luck. In many ways, this is also an area of discovery and constant learning.

How can all of these dynamics be combined to explain a human phenomenon? Let us begin with the phenomena of going to a shop to get a cup of coffee. You have a basic human need for water and nutrition that is very orderly and highly predictable. This is combined in the case of the coffee with the desire for a mildly addictive stimulant. As you leave your home to walk to the coffee shop, you immediately encounter crowd dynamics that may speed or impede your progress to the shops. When you reach your favourite coffee shop, you see that a new coffee shop is open on the opposite corner of the street competing for your business. These shops are engaged in the complex process of competition (very similar to species competition for survival under biotic complexity). In a process of conscious complexity, you are enticed to enter the new shop by its pleasant name, *Vic's*, that reminds you of happy memories from your childhood. As you enter the shop a woman is leaving with a cup of coffee. You open the door for her and say 'good morning'. As she turns to thank you a fly randomly lands on her face, blown there by a turbulent gust of wind from a passing bus. She has a dreadful fear of insects from the stories her grandmother used to tell her as a child and immediately flinches from the touch of the fly. The coffee spills, mostly on your clothes. You return home, change your clothes and make a cup of coffee for yourself. The point of detailing this pursuit of coffee is to demonstrate the remarkable orderly, complex and disorderly processes that are the foundation of most commonplace events in human existence.

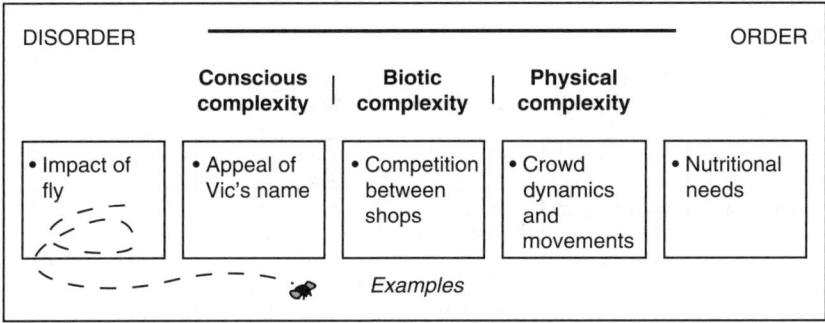

Figure 3.4   The range of complexity in everyday life.

But what if the stakes are higher, when lives are at stake, does complexity still apply? In 1971 Graham Allison, a leading professor of political science at Harvard University, wrote *The Essence of Decision*, one of the greatest English language books in international relations and the best book on the Cuban missile crisis. The basic story is well known. In 1962, responding to the deployment of US nuclear missiles in Turkey, the USSR began secretly deploying missile bases in Cuba. The bases were discovered and a US blockade imposed. The USSR challenged the blockade and threatened nuclear war, but eventually backed down, dismantling the bases in Cuba.

On the surface this would seem to be a simple game of threat and counter-threat that luckily for the lives of hundreds of millions did not go wrong. At one level this is correct. On the other hand, as Allison brilliantly demonstrated, several different political and bureaucratic dynamics both between and within the USSR and USA were going on at the same time. Seemingly rational and irrational strategies emerge from the interplay of these dynamics. For example, when the Soviets were building the missile bases, they built them out in the open and in the same pattern as their bases in the USSR, making them easy to detect by US spy planes, a clear strategic blunder. This was not caused by military stupidity or poor implementation, but caused by the centralised control over Soviet military engineers. The engineers were told to build missile bases in Cuba. They had a model from the USSR, in the open and in a certain pattern and they did as they were told. On the US side, the decision to form naval blockade to stop the Soviets shipping the missiles to Cuba was fraught with military, bureaucratic and personal rivalries. In the end it may have come down to President Kennedy's personal naval experience that led him to choose a naval option. Overall, as Allison points out, these different dynamics could explain parts of the crisis, but none explained all of it. As President John F. Kennedy said after the crisis:

> The essence of ultimate decision remains impenetrable to the observer – often, indeed, to the decider himself ... There will always be the dark and tangled

60 *Tools of complexity*

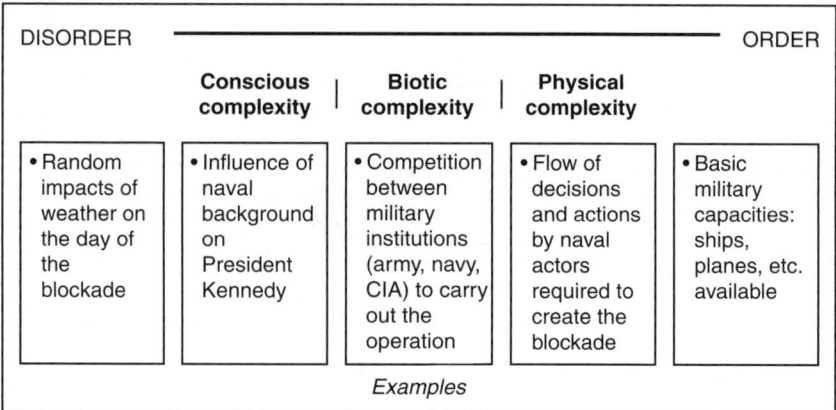

Figure 3.5 Complexity map of President Kennedy's decision to use a naval blockade to stop further Soviet missile development in Cuba.

stretches in the decision-making process – mysterious even to those who may be most intimately involved

(Quoted in Allison 1971: vi)

From a complexity mapping perspective Kennedy's decision to propose a naval blockade would have looked something like the following.

As we will see in later chapters, complexity mapping can help to shed light on the multifaceted nature of most complex human processes.

Fitness landscape

The classic mental model from an orderly perspective is the standard two dimensional x–y graph. The history of the graph goes back several centuries and passes through several cultures. With an x–y graph, any point on a two-dimensional surface can be plotted. For orderly based thinkers, ideally all human systems should be capable of being plotted on an x–y graph and policy actors are strongly encouraged to visualise their policy problems in x–y graph terms.

For example, whether you are dealing with a poorly performing developing country, inner city school or diabetic patient, the approach to all of them is fundamentally the same. First, get them to move from point A to point B as quickly as possible and then keep them moving along from point B to point C for the foreseeable future. From this perspective, the goal is to get to point B as fast as possible and to stay stable at that level for as long as possible. This type of perspective is exactly what many policy actors want to see because it is easy to follow the progress of any given country, institution and/or individual (where are they on the graph?), easy to calculate where the average country/institution/patient is on the graph, and easy to judge how efficient any given policy/treatment is

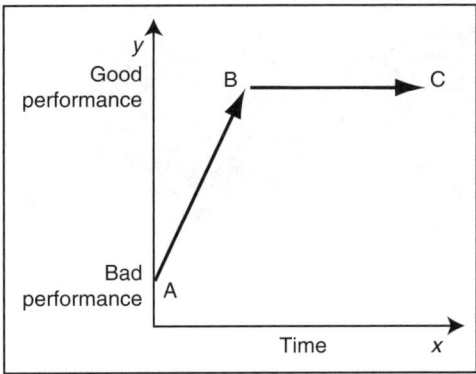

Figure 3.6 Typical *x–y* graph.

(merely compare where all of the comparable countries/institutions/patients are in relation to the average position or in relation to set targets). All of this should sound very familiar since it is the dominant policy framework. However, this perspective has several disturbing implications.

- There is an endpoint B. The goal is to get there fast and stay there. Any strategy that gets them to B faster must be better. Any strategy that keeps patients/institutions/countries on the B to C line longer must be better.
- Experts know how to get patients/institutions/countries from A to B. They have seen many others get there so the key is to do what the expert says (what others have done) and get to B as fast as possible. Once they are at B they are encouraged to do exactly as they have done before as this is the most likely strategy for keeping them on the line from B to C. Emulation of previous successful strategies is the key. Experimentation, exploration and learning are best avoided and left only as a last resort.
- Poorly performing patients/institutions/countries, due to their lack of knowledge, must be fundamentally passive, rely on experts and do as they are told.
- In general, failure to move from A to B quickly or falling off the B to C line implies a failure to follow the rules in the patient/institution/country. It is intrinsically the patient/institution/country's fault. Experts are infallible.

A different point of view: the fitness landscape

For simple orderly phenomena, the *x–y* graph is an excellent tool. To calculate the movements of billiard balls on a table, the trajectory of missiles, and so on, the *x–y* graph is all you need. However, for complex systems you need a way of modelling and imagining a system that can move in varying and unpredictable ways over time. You need a way to show the probability of a system to move

62 *Tools of complexity*

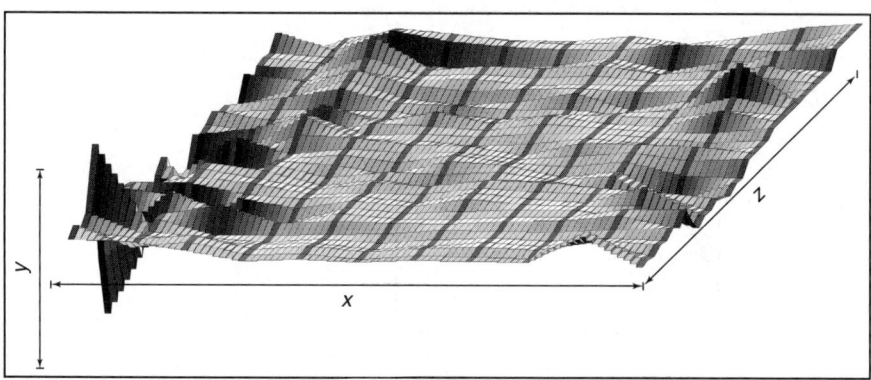

Figure 3.7 Example of a fitness landscape.

in a multitude of directions, a way to show how the system 'fits' with varying situations and circumstances. You need a fitness landscape.

Fitness landscapes were originally designed by population biologists and used in a variety of modelling applications (Coveney and Highfield 1995; Kauffman 1995), and are mathematically represented on a three dimensional graph with x, y and z axis as in Figure 3.7.

In this picture, valleys represent areas of poor fitness, mountains represent high fitness and flatlands represent areas of neutral fitness.[3] Time can be an important factor in fitness landscapes. For example, if the z axis represents time, then one can graphically represent how an actor/unit has responded to their evolving environment over time experiences. We will see how this works in later chapters. Likewise, fitness landscapes are excellent for capturing the symbiotic relationship between multiple interacting actors/units. For example, complex systems such as schools of fish, herds of buffalo or entire species, are constantly moving through an evolving fitness landscape where new predators, food sources and numerous other factors combine to influence their chances for survival. Both the fitness landscape and the complex system change and adapt continuously. A 'fit' actor (which may be a complex system in itself) might find itself in a position of low fitness (due to the activities of competitors that change the topography of the landscape for the overarching complex system) and unless it adapts it might find itself at a great disadvantage for a lengthy period of time or dead. In other words, the fitness landscape may look and feel different for each actor within the system – one set of conditions is all mountains to one actor, by valleys to another. Some of the basic rules for survival on a fitness landscape are adaptability, flexibility, learning and balance. As long as the 'system' survives the fitness landscape continues to evolve through time like a never-ending conveyor belt.

It doesn't take much of an imaginary leap to conceptualise how fitness landscapes would relate to individual human situations. To begin, imagine a landscape that is full of flatlands, valleys and mountains that stretches endlessly

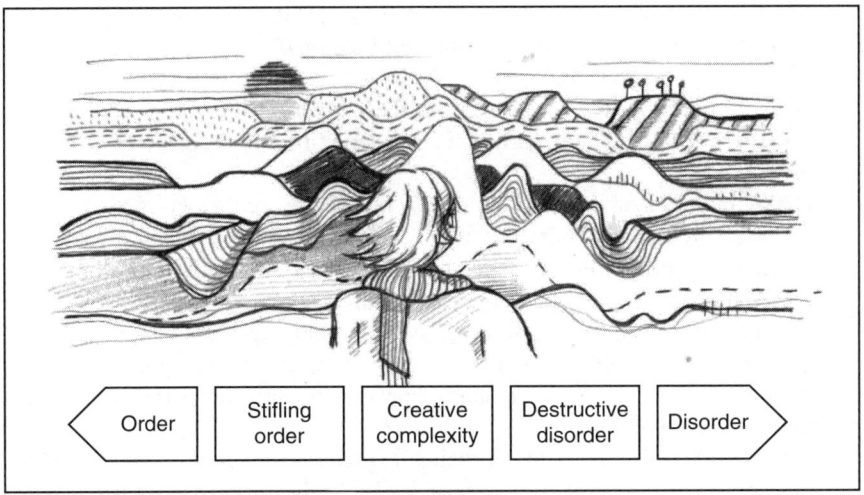

Figure 3.8 Fitness landscape combined with balance and range of outcomes.

into the future. Now, imagine that the valleys represent zones of poor performance, the mountains are zones of good performance and the flatlands are areas of neutral performance. In this landscape a number of mountains, flatlands and occasional valleys are clustered straight ahead of you. On either side, the mountains begin to shrink and become less common, the flatlands are fewer, and the valleys are more numerous and deeper. If we combine this with our concepts of balance and range of outcomes we can generate a good image of the fitness landscape for many complex situations.

Straight ahead is the path of reasonable complex strategies: an evolving general framework, significant degree of local autonomy, high level of local connectivity, and so on. This zone has many mountains of good fitness, but these don't stay the same all of the time. They are clustered in the reasonable strategies zone because these strategies are the most likely to produce good balance. However, even in this area hidden valleys abound (such as shock events or new challenges) that a strict adherence to previous strategies cannot solve. On either side of the zone of reasonable strategies lie increasingly uncertain areas of deep and frequent valleys and only occasional mountains. As this image implies, complex systems struggle in these areas. On the order side, rigid frameworks and structures may succeed for brief periods of time but are incapable of dealing with new challenges or threats. Similarly, on the disorderly side, lacking a stable framework or structure the system will lurch from one position to another, unable to establish healthy stable patterns. We will explore this in more detail in Chapter 5.

Whether it is a patient, institution or country evolving through time, the primary tactics of these complex systems travelling through their fitness landscape are adaptability, flexibility, learning and balance. As opposed to our orderly vision of a walk to the mountaintop of control, a trek through the fitness landscape reveals a number of remarkable 'common sense' implications:

- There is no endpoint to a fitness landscape nor is there any final resting point. The primary goal and tactic is adaptation and balance to changing circumstances.
- The main actors on treks on the fitness landscape are the actors themselves. They are actively moving through the landscape and making essential choices that change the landscape for themselves and others. Their opinions, experiences and learning matter. Experts are secondary.
- Learning is essential and it never stops. Being aware, making choices, experimenting, exploring is how an individual, institution or country learn about their particular evolving landscape and develop tactics for dealing with the inevitable unknown.
- Mistakes, mis-directions and occasional stumbles into valleys are NORMAL and not a mark of weakness or failure. The only mistake is refusing to see them as normal and not learning from them.
- Change is normal. Desperately repeating the tactics of the past (even if they were successful) is no guarantee of future success.

In this book we will not be using the fitness landscape as a modelling tool, but as a visual metaphor for moving away from mechanical perspectives and metaphors. Nevertheless, as we will see in later chapters, these implications lead to very different policy strategies than those suggested by the traditional orderly perspective.

The Stacey diagram

One of the most powerful tools of complexity is a simple diagram developed by Professor Ralph Stacey to explore the nature of decision-making in complex situations (Stacey 1993, 1996, 2000). To build a Stacey diagram you start with an x and y dimension. The horizontal dimension is the 'certainty axis' (close to certainty/far from certainty about outcomes) and the vertical dimension is the 'agreement axis' (close to agreement/far from agreement between the actors). The certainy axis is divided into two ends.

a. *Close to certainty*: at this end, issues or decisions are close to certainty when cause and effect are known. The problem is generally reducible and fully understood and clear techno-rational decision-making can be applied.
b. *Far from certainty*: at the other end are decisions that are far from certainty. These situations are often unique, highly unpredictable and the actors involved have no prior experiences to use to predict outcomes. Cause and effect are unclear.

Tools of complexity   65

The agreement axis is also divided into two ends:

a. *Close to agreement*: these are decisions where all of the actors involved agree on the nature and type of decision.
b. *Far from agreement*: in this area, there is little agreement between the actors on the nature and type of decision.

By bringing these dimensions together Stacey highlighted five different zones which he illustrated graphically:

Zone 1: close to agreement, close to certainty
Zone 2: far from agreement, close to certainty
Zone 3: close to agreement, far from certainty
Zone 4: far from agreement, far from certainty
Zone 5: mixed agreement and certainty

These zones are not dissimilar from our five boxes discussed in the complexity mapping tool. In many ways they represent a similar range of processes applied to decision-making. But, what are these zones and how do they relate to complex decision-making?

Zone 1: close to agreement, close to certainty: the zone of the technical expert

This region, where the problem and response are clear and actors agree on how the problem should be solved, is the zone of the technical expert. If you fall and have a simple leg fracture, all experts agree that the leg should be reset in a plaster cast and you are very unlikely to disagree with them. If your country is suffering from

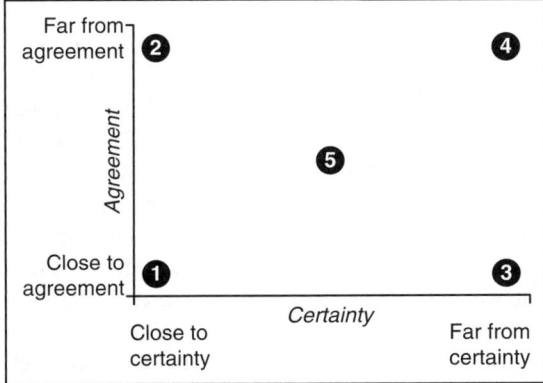

Figure 3.9   Stacey diagram step 1.

starvation, organising immediate food supplies is the agreed solution. In general, this zone tends to be where actions and decisions are based on past experiences, where data from the past can be effectively used to predict actions for the future.

Zone 2: far from agreement, close to certainty: the zone of 'political' decisions

Some issues have a great deal of certainty about how outcomes are created but high levels of disagreement about which outcomes are desirable. These issues are more complicated and are very similar to 'political' issues that are continually debated in society, social groups and even individual families. In this context, local and/or individual views are important. Democratic processes, compromise and negotiation are all paramount. In this zone there is no inherently right answer, but contested political ones.

Zone 3: close to agreement, far from certainty: the zone of judgemental decisions

Some issues have a high level of agreement but not much certainty as to how to make them happen; that is, the favoured outcomes are clear but the means of getting there are not. The cause and effect linkages are therefore unknown. In this zone, the goal is to go towards an agreed-upon future state using judgement even though the specific paths of how to get there are not yet clear. For example, everyone may agree that decreasing obesity in a given population would lead to health improvements, increasing the reading skills of 11-year-olds would improve general education levels or reducing global warming is important. However, the best way to reach these is very uncertain.

Zone 4: far from agreement, far from certainty: the zone of disorder

Looking at the opposite end of the spectrum, some issues have very high levels of uncertainty and disagreement. These situations are the most unpredictable and can be the most difficult to deal with. The traditional methods of planning and negotiation are insufficient and actors often deal with such situations by 'muddling through' or avoidance. Examples include the positions of local actors caught in highly chaotic civil wars or conflict situations, companies caught in turbulent economic times or individuals dealing with particularly erratic chronic health conditions and multiple interacting mental illnesses/disturbances (the destructive cycle of child abuse, self-abuse, abuse of others, etc.). Moving through this zone is, as a leading complexity thinker argued, 'like walking through a maze whose walls rearrange themselves with each step you take' (Gleick 1987: 24).

Obviously, this can be a very dangerous zone of uncertainty. And yet, is this zone so uncommon in our daily lives? For example, choosing life partners is one

Tools of complexity 67

of the most uncertain and 'far from agreement' long-term decisions that can be made. An 'expert' opinion would only be of marginal use. And yet, we do it all the time and mostly reasonable outcomes occur; although about half of all life partnerships now lead to divorce. In this zone, muddling through is about the best we can do. Ignoring and denying the reality of this zone does not make it go away!

Zone 5: mixed certainty/agreement: the zone of complexity and learning

This zone lies between the regions of disorder and traditional management approaches and mirrors elements of the middle boxes of our complexity map. In this region, traditional management approaches are not very effective so it is the zone where people need to learn new approaches, new techniques, and maybe even unlearn traditional methods. This zone requires a range of approaches to deal with complex situations (open agenda building, brainstorming, muddling through, etc.). Some of these approaches will have no evidence base but will be based on expert opinion and/or intuition. Visually, the zones can be combined to make Figure 3.10.

Essentially, what the Stacey diagram demonstrates is that most of our fundamental decisions are based on a healthy combination of orderly, disorderly and complex decision-making. There is nothing inherently superior about technical-rational decision-making. Where it is appropriate it is an excellent tool. Where it is inappropriate it can be a dangerous weapon. Moreover, as we will see in later chapters desperately trying to push all decision-making processes into the

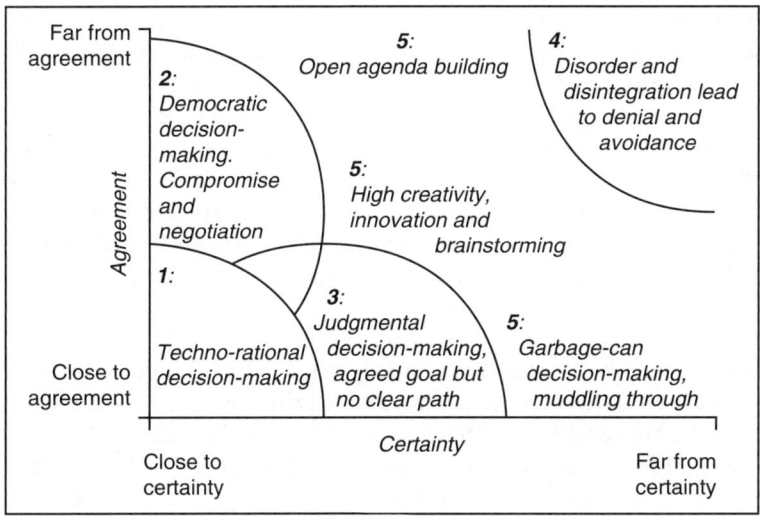

Figure 3.10 Stacey diagram step 2.

orderly area of Zone One is a maladaptive and dangerous practice. And yet, it is the underlying tendency of the traditional framework of policy making. Helping to move policy actors away from that tendency is what this book and the Stacey diagram are all about.

Stakeholder involvement and soft systems methodology

As we have seen elsewhere there is a quantum difference between orderly and complex systems and most human systems exhibit orderly, complex, and disorderly modes of behaviour. These simple, self-evident facts have major consequences when it comes to the formulation and execution of policies in the social, political, and economic arenas. However, as we saw in Chapter 1, from an orderly modernist perspective all systems should behave in an orderly and predictable manner. Command-and-control from the top mediated through hierarchical management structures is seen as the most appropriate style of management, be it on the industrial assembly line or in national and international affairs.

As we have stressed earlier, these criticisms have been made before and were well captured by the work of Rittel and Webber and their work on 'wicked problems' in 1973.[4] Wicked in this context does not mean evil but the kind of problem that has perplexing characteristics that require solutions that are not encountered in 'tame' problems. In this case, even the tags 'problem' and 'solution' present a difficulty. The presence of conscious agents gives these features a fluid nature. What is considered a problem, and in what way is it a problem, depends heavily on who is making that distinction. The same person might view the 'problem' differently over time. And crucially, the 'problem' might change during the process of 'solution'.

The above fluidity makes life uncomfortable for decision-makers and experts sitting at the top of the pyramid. Even if they are saintly in their devotion to the public good they are faced with continuously shifting messages from those they serve as to the nature of 'problems' and solutions'. They are obviously in a difficult situation, and as more complexity is added with the passage of time the more their position becomes untenable. Present-day unhappiness and disillusionment with the 'leaders' in almost all fields is not unexpected.

What is the way out of this dilemma? How does one manage under these shifting conditions? The Australian Government Public Service Commission is in no doubt. In the conclusion to it report on 'wicked problems' the Commission states:

> Many of the most pressing policy challenges for the APS involve tackling wicked problems. Wicked problems are characterised by social complexity – they cross the boundaries of APS agencies, they cross jurisdictional boundaries, stakeholders (and experts) often disagree about the exact nature and causes of the problems and, not surprisingly, they disagree about the best way to tackle them. A key part of the solution to many wicked problems

involves achieving sustained behavioural change. It has become increasingly clear that a disengaged and passive public can be a key barrier, and is a factor in the policy failures around some of Australia's longstanding wicked problems. In the areas of welfare, health, crime, employment, education and the environment, *significant progress requires the active involvement and cooperation of citizens* [emphasis added].

(Australian Government Public Service Commission 2007: 35)

Clearly, at least part of the way forward is through real *stakeholder involvement*. The traditional steep hierarchy that separates stakeholders from decision-makers by layers of bureaucrats and 'experts' is obviously not fit for complex situations. A healthy and continuing dialogue about problems and solutions could not be undertaken sensibly within such an outdated model. Moreover, the Australians are not alone in realising the limits of the current model. The Canadian Institute of Governance organised a conference in 1998 that sought to define a better model based on closer linkage between all actors.[5] A similar theme was picked up by the OECD in 2001 in a report titled 'Citizens as Partners'. (www.oecd.org). In the UK, the Department of Health's Modernisation Agency (now amalgamated with another agency) published thirteen 'Improvement Leaders' Guides' in 2005 based on a systems approach to health. The guides, especially 'Working in Systems', explored the difference between orderly systems and the complex systems regularly found in all aspects of the health sector and then went on to set out in some detail the changes in working practices and culture required to successfully tackle the wicked problems encountered. However, as demonstrated by the continued dominance of the 'audit' and targeting culture in most policy areas, it is far from certain that governments are fully willing and able to put this into practice.

In addition to active stakeholder involvement, dealing with conscious complex systems requires a new form of *soft systems methodology* (Wilson 2001). In orderly systems 'problems' and 'solutions' are separate activates. A defined problem of sending a human being to the moon, for instance, is obviously followed by a separate phase of defining a way forward, which is again followed by a separate implementation stage. There is a start and finish to this sequential process that flows naturally through logical separate steps. This management style is often referred to as the 'waterfall model' in project management and software development and is shown in Figure 3.11.

However, when the perception of the problem might be influenced by time, personal choices, and by the solution adopted to solve the problem as originally specified, the process becomes infinitely more fluid, as demonstrated by Figure 3.12.

Clearly, both problems and solutions have to be revisited over and over again as time goes on. The 'waterfall model', which has been the traditional way of attacking all problems, is clearly inappropriate in situations embracing the activities of conscious complex systems. To deal with those systems, a new model

70 *Tools of complexity*

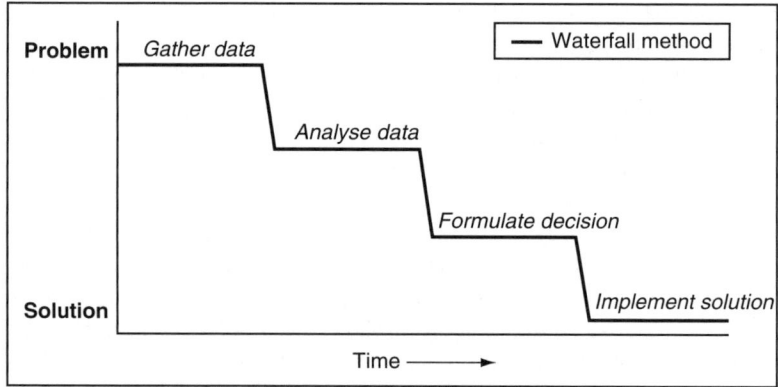

Figure 3.11 Traditional decision-making process. Based on an original image at http://www.metronetiq.com/archives/2007/09.

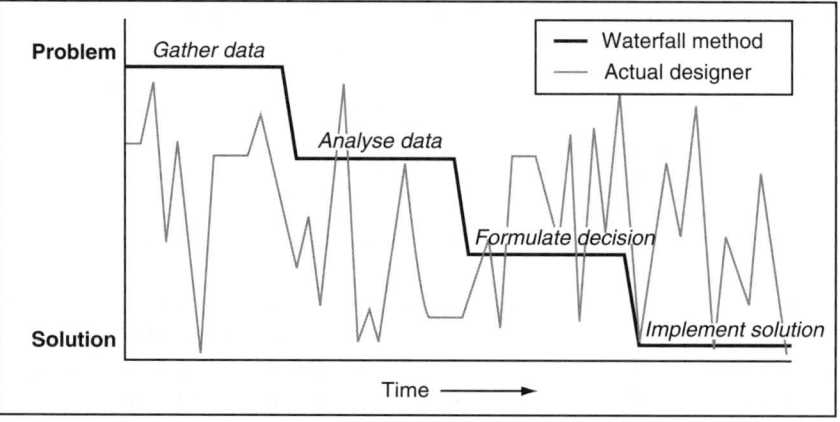

Figure 3.12 Traditional decision-making overlaid with real decision-making process based on an original image at http://www.metronetiq.com/archives/2007/09.

is needed that can deal with stakeholder involvement and the reiterative processes that recycles through all stages of the work.

Active and continual stakeholder involvement in a reiterative, learning process for human activity systems fits neatly into what has become known as 'soft systems methodology' (SSM).[6] The technique evolved in response to dissatisfaction with practices such as systems engineering when applied to complex systems and 'wicked' problems.

The principal advice in SSM is for those involved in decision-making not to jump to conclusions too early. The process itself is seen as being just as important

Tools of complexity 71

as a means for exploration and learning for all stakeholders as the actual result of the exercise. Clearly SSM could not be described adequately here, but the joint action taken by the participants could be summarised as follows:

- Open exploration of the problematical situation. Sometimes techniques such as SWOT (strengths, weaknesses, opportunities, and threats) and PESTLE (political, economic, social, technical, legal and environmental) are deployed to guide, rather than restrict, discussions. So-called 'rich picture' are also used to present information (about the problematical situation) pictorially.
- Clarification of stakeholders. CATWOE (customers or clients, actors or providers, transformation process, worldview or 'climate', owners or funders, and environment or constraints).
- 'Root definition' tries to sum up what has been learnt so far to gain an understanding of what is the fundamental purpose of the system under consideration. Why is it there?
- A 'conceptual model' (including resources required and service criteria that would indicate success or failure) that describes the ideal model that would deliver the job outlined in root definition. At this stage the participants try to forget about the existing structure.

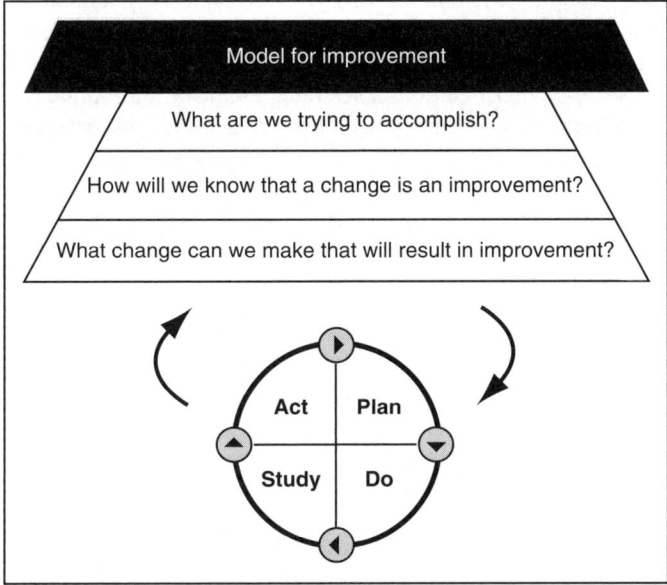

Figure 3.13 SSM inspired organisational strategy plan. Used in the NHS improvement Leaders' Guide, *Working in Systems (2005)* but originally published in langley G, Nolan K, Norman C, Provost L, (1996). The Improvement Guide: a practical approach to enhancing organisational performance, Jossey Bass Publishers, San Francisco.

We included a very brief outline of SSM above simply to make a few key points. First, management of conscious complex systems is a social activity and is not confined to a few 'bosses' supported by 'experts'. Second, such a process could only proceed in 'shallow hierarchies' in which persons at all levels have a key role to play in managing the organisation. This raises interesting questions about risk sharing and risk taking. Third, learning is a critical factor for success. As stressed in the preceding chapter, management of human activity systems, like learning, is a process of continual exploration. There is no beginning or end.

The last point reveals a feature that is of cardinal importance. Management of 'wicked' situations encountered in conscious complex systems is an open-ended activity. The aim is to define a desirable direction of travel, at a given point, and then seek to nudge the system in that direction through a cyclical process of learning, monitoring and adjustment. The adjustment may well include a revision to the direction of travel itself. Such SSM strategies are now widely available and can be visually represented in the manner shown in Figure 3.13.

The contrast between the above process and that applied to tackle hard systems could not be clearer. Just as we saw in the Stacey diagram, some human processes fit with hard systems management and others do not. A complexity perspective is essential in helping one to recognise this and choose the right methods for responding to the particular systems/problems/solutions. In the next chapters, we have selected policy areas that illustrate the inadequacies of jumping to the wrong conclusion by treating complex systems as orderly, mechanistic phenomena. The astonishing point we seek to expose is the lengths to which decision-makers have gone to ignore the shortcomings of current practices, the subsequent enormous costs of such orderly interpretations and strategies and how a complexity perspective and some of its tools could have been used to avoid or correct these failings.

4 Politics

> Any academic discipline that wants a place at the trough, but is unable to offer the predictions and the technology provided by the natural sciences, must either pretend to imitate science or find some way of obtaining 'cognitive status' without the necessity of discovering facts.
>
> (Richard Rorty 1991: 35)

If economics is the 'dismal science' then politics must be the miserable science. With a huge, diverse and constantly changing field to cover, working away without major rewards (no Nobel Prize!), struggling to prove that you are relevant and trying not to be seen as captured by a particular interest group or ideological position, one should spare a thought for the long-suffering political scientist. As the above quotation demonstrates, it is no wonder that the dominant politics journals position themselves within the orderly paradigm or that many academics cling to a vision or strategy of order that can give them the epistemological and methodological foundations (and respect and fiscal remuneration) of the 'hard' sciences.

Obviously, not all political scientists are enthralled by the orderly paradigm and there is a mountain of alternative journals to choose from. Our point is that despite the commonsensical complexity of politics and the undeniable evidence of divisions within the discipline (Left *versus* Right, modernist *versus* postmodernist, etc.), it remains embedded in the orderly framework.[1]

To begin to chip away at this dominance and to explore what complexity has to offer in this field, this chapter will begin by using the 'complexity mapping' tool to delve into some of the basic implications of complexity thinking for politics in general. Following this, it will examine how complexity reshapes the basic understandings, regularities and threats to democracy. Next, it will critically explore one of the most influential ideas in recent political debates, New Labour's concept of the 'third way'. Finally, keeping with the theme of 'thirds', the chapter will turn to debates surround the role and impact of civil organisations between the market and the state or the so-called 'third sector'. After a brief discussion of the third sector and its relationship to the EU we will use the 'balance and range of outcome' tool to consider what the EU should do to support the third sector at the European level.

What would a 'complexity map' of politics look like?

Politics, like other intellectual disciplines, easily fits into a complexity framework. It is just a matter of inserting the particular aspect that you want to study. For example, if we apply a complexity map to the general concepts of politics we get something like Figure 4.1.

Moving from Left to Right, a good example of the disorderly aspects of politics can be found in the particular nature of local aspects of the long-term development of the political system. For example, in the American political system, who could have guessed in the late 1960s that a popular president and lifetime politician, Richard Nixon, could be brought down by a chain of events arising from a bungled burglary in a Washington hotel. Likewise, who could have predicted that a telephone call from 'that woman' would almost bring down the presidency of William Clinton in the 1990s. Similarly, though the Electoral College has overridden the popular vote before, who could have predicted that it would do so again under such controversial circumstances during the 2000 presidential race and bring the debate of the 'hanging chad' to a worldwide audience. The obvious vagaries and unintended outcomes of other political systems (the current crisis of MPs expenses in the UK, sexual scandals of Berlusconi in Italy, etc.) are too numerous to mention.

In terms of conscious complexity, there are a multitude of obvious examples. Just look at the different interpretations of socialism or conservatism in different national and regional contexts. In the US, socialism has virtually no resonance and is only vaguely associated with old-style communism despite the fact that the US had a thriving socialist party at the beginning of the twentieth century. In France, modern socialism emerged out of a strategy for political power on the Left and to isolate the French Communist Party. In Spain and Greece, it was a

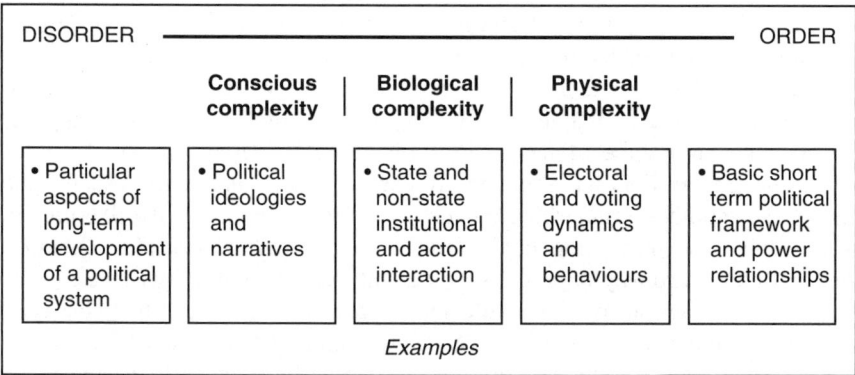

Figure 4.1 The range of complexity in politics.

force for democratisation in the 1970s and 1980s. Likewise with conservatism, in some parts of the US it is equated with opposition to the federal government and the protection of the right to own handguns. In other parts it is the defence of a Christian political order. Meanwhile, in the UK conservatism supports a traditional sense of elitism and deference. In other countries it stands for a militant defence of free markets. All of this ignores the widely divergent meanings of these words in the non-Western world where, for example, socialism was often equated with anti-colonialism.

Dynamics of biotic complexity can easily be found in the evolving actions of political institutions, such as political parties, interest groups or governmental bureaucracies as they struggle for scarce resources like votes, influence or funding. These major political actors are caught in a distinctive symbiotic environment in which they respond to the actions of others in a context of fundamental rules and extensive regularities (elections, legal rules and norms, bureaucratic practice, etc.). Survival and success are specific to each situation. For example, the strategies of a successful political party in Spain might be useful to a party in Norway, but only if translated through the prism of the local situation. The field of comparative politics has struggled with this limitation throughout the twentieth century. Direct implantation of political institutions, as we will see in later chapters, is a virtual guarantee of institutional failure.

Likewise, within these systems there are aspects that mimic physical complexity. Elections have generally clear rules, democratic legislatures run on obvious electoral lines and bureaucracies often have patent functional rules. With all of these rules and orderly structures, quantitative and reductionist methodologies begin to be more and more applicable to these phenomena. A quick glance at some of the dominant political science journals, particularly the *American Political Studies Review*, shows an abundance of studies on rational decision-making and electoral outcomes. However, despite all of this regularity, there is significant room for complex outcomes, the unpredictable political gaffe that leads to an electoral landslide or accidental bureaucratic outcome that causes a policy scandal and destabilises a traditional political coalition are two general examples. Moreover, like explaining the weather, these rules and regularities create higher probabilities and greater predictive capacity, but do not actually explain the multitudinous factors behind any given bureaucratic and electoral decision.

Lastly is the realm of order. Here our closest analogy in the political world is the short-term regularities of most political systems. For example, that there have been modern political parties, interest groups and bureaucracies throughout the twentieth century and earlier forms of these organisations for long periods of time clearly indicates that one should see them as fundamental regularities, particularly in the short-term. One can predict with almost law-like certainty that the Labour and Conservative parties will continue to exist and dominate the British political landscape for the next 6 months. However, will they continue to do so in the next 2–5 years, 5–10 years, 10–20 years? Obviously, uncertainty grows with time. The division of societies into elites and non-elites is another example of a fundamental regularity that could be considered, in the short term, as an orderly phenomena. However, once again, the particular aspects of that regularity can be extremely variable with place and time. But these generalisations are the easy part. How does complexity relate to particular political concepts like democracy or the third way?

Complexity and democracy

At first glance, one may think that the Newtonian foundation of the social sciences would have little to do with liberal democracy. Democracy as a concept is neither simple nor clear. It is made even more obscure because it has been adopted over the ages as a slogan by all kinds of political movements, businesses, interest groups, and individuals. On the face of it, the Western liberal meaning of the word is easy to grasp; a country is democratic when the people are free to choose their own form of government and to elect their representatives at regular intervals. In general, modern liberal democracy offers personal freedoms, equality before the law, and universal suffrage. The Declaration of Independence adopted by the American Continental Congress in 1776 and the Declaration of the Rights of Man and the Citizen approved by the French National Assembly in 1789 are often quoted as sources for the main features of Western democracy. The Atlantic Charter, agreed in 1941 by the US and Britain and later embraced by most countries, led to the formation of the United Nations in 1945 and provided a modern interpretation of what is understood by democracy at the national and international levels.

However, in some respects the above features are sometimes as flawed in practice as the ancient concepts of Greek and Roman democracy. It is easy to describe what democracy might entail but adoption of the principles and applying them broadly to involve all facets of the life of individuals, businesses and nations follows a lengthy, uncertain, and tortuous route. For instance, a constitution dating back to 1874, and amended regularly after that date, allowed women in Switzerland to have a voice in choosing their government only as recently as 1971. Equally, the American Declaration of Independence stated categorically that 'all men are created equal', but the poor were barred from voting until the late nineteenth century, women had to wait until the early twentieth century, while black people were excluded from that lofty aim up to the second half of the twentieth century.

Nevertheless, a core Newtonian foundation can be found for the current structure of Western liberal democracy. A key work that elucidates this foundation is Benjamin Barber's *Strong Democracy: Participatory Politics for a New Age* (1984). Arguing from a pre-complexity framework and drawing heavily on the American philosopher John Dewey (particularly his seminal *The Quest for Certainty* (1960)), Barber convincingly demonstrated that many liberal democratic theorists' pre-conceptual and epistemological frameworks were entirely Newtonian. As Barber points out, in an attempt to duplicate the success of the Newtonian method, both classical (Descartes, Hobbes, Locke) and modern liberal theorists (John Rawls, Robert Nozick) searched for firm and certain foundations to their theories and then attempted to build logical structures upon these foundations.

These foundations were usually founded in the belief in self-interested atomistic individuals rationally pursing their distinctive self-interests in a context of other individuals pursuing their own interests. Given this base, noticeably similar to Newton's theory of particles in motion, a full-scale theoretical construction of social and political relations was created, neatly mirroring the physical sciences. As Barber demonstrates with the concepts of liberal freedom and power:

> Rendering freedom and power in physical terms not only misconstrues them, it produces a conception of political liberty as entirely passive. Freedom is associated with the unperturbedness of the inertial body, with the motionlessness of the inertial frame itself ... The modern liberal appears to regard it as a republican ideal: man at rest, inactive, nonparticipating, isolated, uninterfered with, privatized, and thus free.
>
> (Barber 1984: 36)

The fundamental problem for liberal democratic theory is its quest for a certain, knowable and universal base. If this base could be found, like the laws of gravity for Newton, then the fundamental aspects of society could be deduced. However, given the orderly, complex and disorderly development of human history, this search would not be a fruitful one. Moreover, it tends to lead the theorist towards a rigid, inflexible and dogmatic view of human nature and social order. For example, once liberal democrats have found their certain foundation they tend to concentrate on the implications and logic of their initial positions rather than the complex reality of human interaction. Again, as Barber succinctly argues:

> The quest for certainty in political thinking seems more likely to breed orthodoxy than to nurture truth and in practice tends to promote the domination of method over substance ... In an attempt to mimic the hard sciences, of which they rarely have a true understanding, these social scientists have tried to subordinate every understanding of reality to some orthodox construction of understanding.
>
> (Barber 1984: 47)

What would a complexity theory of democracy look like?

Democracy, like politics, easily fits within a complexity framework, displaying orderly to disorderly aspects and dynamics. Complexity would emphasise that democracy is not a single phenomena or end-point, but an evolving process that is distinctive in every area where it has emerged. Some of its global patterns are predictable (competing interests, free elections, etc.) but useful interventions for promoting democracy are restricted to enabling interactions to proceed in a manner that produces self-organised stable patterns rather than disorder. In this instance local freedom of action, learning, and access to information are vitally important, as control is limited to observation of outputs and 'encouragement' for the system to evolve towards generally desirable ends.

Promoting democracy is, like promoting other complex concepts, a reiterative process that relies on slow, and uncertain, progression. From a complexity perspective imposing democratic institutions on a country and expecting them to work within 6–12 months (as in the case of Iraq) is absurd. Democracy requires time and a delicate balance between variety of local interactions and global stability that allows the system to survive through continual adaptation, all the while steering an uncertain path between stifling order (dictatorship) and debilitating disorder (civil breakdown/war). Under these conditions, minor variations (a new party in power, new electoral rules, etc.) cause some change but this is normally contained within the overall pattern, or *attractor*, existing at the time. Only occasionally do these changes succeed in triggering a major shift, punctuated equilibrium, into a new pattern (from a bi-party to multi-party system, from unitary to federal government, from dictatorship to democracy). Evolution of democratic practices, as in all other complex processes, does not lead to an optimal end state. It is 'an open-ended process by which a structure evolves through interaction with its environment to deliver a better performance' by means of small effective improvements (Coveney and Highfield 1995: 118). A 'good enough' democracy which combines some form of stable fundamental framework and as much local freedom and diversity as possible seems to be the sum total of what a healthy dynamic democracy should strive for.

Fundamental regularities and threats

There are two significant *regularities* in complexity theory that are directly relevant to democracy. First, average complexity tends to increase. And second, entities with higher complexity stand a good chance of making the most gain in new situations. The ubiquitous presence of elites and hierarchies in businesses, nations and within the interstate world system, the emergence of hegemonic powers and the inclination for the rich to get richer hint at political economic similarities. Time piles complexity over even more complexity. Culture, society, economics, politics and democratic practices tend towards increasing complexity in the same manner, and as elsewhere it takes time for complexity to build up. Layers of interconnections and adaptations have to come into being at a pace that could not be rushed or acquired through higher spending, imitation, external pressure or imported expertise,[2] and the process has no end.

As individuals, businesses, and nations compete and cooperate in the search for survival and dominance, a hierarchy is an inevitable consequence, or *frozen accident*.[3] The rigidity and shape of the hierarchy in the context of democracy is determined by a number of fundamental and local factors, including history, tradition and form of government. In undemocratic situations, the hierarchy is generally steep and political and economic interactions are monopolised by a few people. Order is achieved at a high price involving slow, or even negative, evolution. Under democratic conditions, by contrast, the hierarchy is distinctly flatter, the elite embraces more individuals, and most people are reasonably free to interact locally in a huge variety of ways. Here, healthy evolution is not only possible but also probable. The global hierarchy comprised of a hegemonic power, leading nations, and others, follows similar traits. Speedier progress is more likely when that hierarchy is less steep and where local variety is not seen as a 'threat to world order'.

Cascading rewards and penalties, structured around the needs of the elites at individual, national and international levels, are deployed to keep the hierarchical system stable. For instance, chairmen of companies ostensibly recruit non-executive directors to their boards to foster good governance. But, non-executive directors also set salaries for chairmen. In practice, a few persons serve as chairmen in one corporation and as non-executive directors on a number of other boards (Herman and Chomsky 1994). Politicians, even in Western democracies, play a similar game of musical chairs: they are either in power or in opposition but even when they are out of power they are readily assimilated into a number of lucrative positions for the duration. In other words, democracy is a matter of degree and style rather than an absolute state. Hence, Western democratic systems have not reached a state of perfection; there is no such end state in any case. Moreover, buffeted by both challenges and opportunities they are continually evolving under changing circumstances.

On the contrary, from an orderly perspective Western democracy is the ideal political form that all other systems should aspire to. The main threat to it is the resistance of non-democratic systems. Hence, wherever possible spread Western democracy and this will shortly transform those societies into like-minded liberal allies. Occasionally, this leads to an 'overactive' democracy where the populace elects unacceptable actors (Chile, Nicaragua, Egypt, Indonesia, Iran, Palestine, etc.), but this merely indicates that the system has not reached the appropriate level and that more control/steering is required.

From a complexity perspective, the main threats to democracy are the continuing belief in and pursuit of an 'end' state for democracy, the internationalisation of orderly models of democracy and economic globalisation and inequality.

Despite the mounting evidence of the dynamic nature of democracy, order-based thinkers continue to try to demonstrate that democracy has reached a final 'end' state, the latest and most famous example is Francis Fukuyama (1992). Ready-wrapped in convincing and possibly inviting garb, democratic theories which argue for an end state of democracy are actually potent threats. Reduction of choice and variety are the biggest dangers in view of the dynamic nature of the processes involved in the emergence and development of democracy. From a complexity perspective, rigid consensus and uniformity could lead to stultifying order as opposed to the healthy tension of creative complexity. Of course, the opposite is true as well, without some agreed fundamental framework creative tensions could spill over into destructive disorder (civil war). The challenge for democracy is to find a continually evolving balance between a fundamental framework and as much local interaction and openness as possible.

Likewise, the internationalisation of orderly models of democracy is an equally dangerous tendency. Pressure from the so-called 'Washington consensus' (Serra and Stiglitz 2008) and their desire to impose precise models of democracy, commerce and industry undermines the variety and choice as well as checks and balances of emerging democracies, crippling their ability to survive and prosper. External pressures from international organisations (the IMF/World Bank) or dominant national elites (the US State Department or UK Foreign and Commonwealth Office) aimed at constraining local choice and freedom of action provide the most inhospitable conditions for the dynamic process of democracy to gather momentum. Moreover, whether well- or ill-intended they are often perceived as a threat to the emergence and progressive evolution of democratic practices and often lead to sham 'donor democracies'.[4]

Lastly despite the recent economic downturn, with the growth of economic globalisation and inequality, power is shifting away from national governments to larger political alliances and vast corporations far removed from the local grassroots level (Klein 2007). This process has led to the surrendering of many welfare reforms introduced after much thought and effort and over a long period of time (Deacon 1997). These rights and reforms, such as the basic rights to access education and health services, to name only two examples, are as essential components of any democratic system as political institutions and voting rights (Skogly 2006). Moreover, this decline in general democratic/welfare rights often goes hand in hand with high levels of income inequality which are equally dangerous to democratic systems (Collier 1998 and Horowitz 1998). Hence, from a complexity perspective, this growing inequality fundamentally weakens both the underlying framework (the basic social and welfare rules that help to maintain social cohesion within all societies) and the 'connections' (the level and types of interactions; see Putnam 2001) between the elements (individuals/groups) within the national systems, making the *fitness landscape* of democratic development significantly rockier.

Complexity and the 'third way'

In the final decade of the twentieth century the Left in the US and Europe were in a quandary. Communism was in full retreat. Globalisation appeared rampant. And all they seemed to stand for was a static defence of the welfare state. Out of this turmoil emerged a new strategy: the 'third way'. Based on the work of the esteemed British philosopher and sociologist Anthony Giddens and embraced by President Clinton, Prime Minster Blair and Chancellor Schroeder, the 'third way' promised to revitalise the Left, revive its electoral fortunes and renew the 'future-oriented radicalism' that had been the hallmark of the Left throughout much of the twentieth century. Not surprisingly, the third way, as a concept, was adopted and co-opted in a variety of ways, but its general influence remained unquestionable while its influence continues to reverberate in a number of countries to this day.

The foundation of the third way can be found in the major works of Anthony Giddens. In *Beyond Left and Right: The Future of Radical Politics* (1994) he developed the concept of 'manufactured risk/uncertainty'. Following in the footsteps of a number of post-modern authors and sociological traditions, he argued that, 'the world we live in today is not one subject to tight human mastery' (Giddens 1994: 4). This is not due to the complexity of nature or the failure of humankind to master it, but due to 'our conscious intrusion into our own history and our interventions into nature' (Giddens 1994: 78). For Giddens, due to the development of industrialisation, globalisation, a post-traditional social order and the rise of 'social reflexivity' (Beck 1992), uncertainty and risk can no longer be externalised to natural occurrences, but are increasingly a product of human action and awareness. In other words, despite the hopes of the Enlightenment, where humanity would increasingly entrench its dominance over natural and social conditions, the very pursuit of this dominance had led to the creation of human-manufactured uncertainty and risk which altered the fundamental framework of our relationship to the natural world and human social development.

For Giddens, the early appeal and fundamental weakness of Marxism was that it rested on the belief that problems created by humanity must be resolvable by humans and, consequently, that it had found the fundamental direction of human evolution. This basic belief in the ability of revolutionary (later, technocratic) elites to direct and understand human economic and social interaction manifested itself in the planned economies under communism and the belief in planning in the advanced industrial economies in the early post-WWII period. This model was 'reasonably effective as a means of generating economic development in conditions of simple modernization' (Giddens 1994: 66). However, as societies and economies become more complicated and reflexive, the model became increasingly dysfunctional.

With the collapse of the Soviet Union in 1989 and the difficulties of social democracy in Sweden in the late 1980s, often seen as the most developed social democratic society, socialism lost its radical edge and fell back on a static defence of the welfare state. Meanwhile, neo-liberals, espousing the ability of the unfettered market to solve both economic and social ills, 'appropriated the future-oriented radicalism which was once the hallmark of the bolder forms of socialist thinking' (Giddens 1994: 73).

In order to bring radicalism back to the Left he explored the implications of manufactured risk/uncertainty on a variety of policy arenas, particularly the welfare state. He argued that the Left had to abandon its traditional defence of the welfare state and replace them with ideas of 'positive welfare'. Like Keynesian economic ideas, the welfare state performed well under conditions of simple modernisation and clear external risks. However, with the growth of globalisation and social reflexivity the traditional welfare state became dysfunctional. The attempt by the welfare state to combat stable external risks led to contradictory outcomes which ultimately doomed the Left's defence of the traditional welfare state. He came to the conclusion that without abandoning the earlier belief that history has a particular direction, accepting the new conditions of manufactured risk/uncertainty and reforming its thinking, particularly in regards to its last defensive stronghold (the welfare state), the Left will never regain its historical radicalism.

Giddens later brought these ideas together in his remarkably wide selling book, *The Third Way* (1998). Whereas the Left focused on the state and the Right on the market, the third way argued that it is necessary, 'to go beyond those on the Right "who say the government is the enemy," and those on the Left "who say government is the answer"' (Giddens 1998: 70). Likewise, the third way advocated a 'new mixed economy' that promoted a 'synergy between public and private sectors, utilizing the dynamism of markets but with the public interest in mind' (Giddens 1998: 100). Regarding the welfare state, he argued that it needed to be transformed into a 'social investment state' that would be 'dynamic and responsive to wider social trends', 'promote risk-taking by individuals' and 'invest in human capital wherever possible' (Giddens 1998: 117).

Following the success of *The Third Way* came a torrent of comment and criticism. Some argued that the third way was an 'amorphous' political project (Faux 1999), others that it was merely the reassertion of an earlier British tradition of New Liberalism (Ryan 1999; Driver and Martell 2000). Traditional Marxists argued that it was 'primarily a rationalization for political compromise between left and right, in which the left moves closer to the right' (Hall 1998; Lafointaine 1998; cited in Giddens 2000: 11) and therefore fundamentally conservative. Academics and politicians in Continental Europe argued that it was fundamentally an Anglo-Saxon project and 'was of little use to societies that are further along the road to social justice and more comprehensive welfare provision' (Levy 1999; Lightfoot 1999; cited in Giddens 2000: 24).

Basically, all of these critics wanted the third way to be more specific, orderly, and predictive. Whether arguing that the third way was not distinctive enough from historical precedents, was too vague, was a rationalisation and/or did not fit their particular models, all of these critics were trying to force Giddens to justify his more post-modern, contingent and reflexive framework. Unfortunately, instead of maintaining his uncertain complex position, Giddens responded to these criticisms from a predominantly deterministic and orderly position, namely that the third way was a new type of order that could be opposed to earlier forms of order and was an improvement on these earlier forms.

For example, in responding to his continental critics he argued that in spite of the distinctiveness of most European socialist parties and welfare state regimes there is 'a single broad stream of third way thinking, to which the various parties and governments are contributing' (Giddens 2000: 31). Hence, despite their differences they are all part of a similar linear trend or process. The US and UK are at the front of this process because they 'experienced neoliberal government in "full-blooded form"' (Giddens 2000: 32). This made the Left in the US and UK more willing to question traditional orthodoxies and shift towards the third way. Moreover, argued Giddens, those social democratic countries who appear to be the most traditional, particularly Scandinavia, 'are likely to be more vulnerable to the changes now happening than countries "further back" on the welfare scale' (Giddens 2000: 34). Once again, a developmental line had been drawn, it was just a matter of positioning the various countries along the line.

A few early critics pointed out that the third way 'describes the present in epochal terms, implying that there is only one right way to understand and respond to the real changes occurring in our world' (Rose 1999: 490). Consequently, they (referring to Giddens and Will Hutton, the influential editor of the *Observer* newspaper):

> draw a diagram of history in which a single axis of time catches up all corners of the globe into its current and drags all along its wake ... (all must) become modern or face the destiny of the obsolete – the scrap heap of history.
> (Rose 1999: 471; see also Benton 1999: 44)

Most influentially, Ralf Dahrendorf attacked the 'authoritarian streak' in the third way:

> The great liberation of the revolution of 1989 was that it ended the dominance of ideological thought systems. There are no longer even first, second and third worlds, only varieties of attempts to cope with economic, social and political needs. The Third Way presupposes a more Hegelian view of the world. It forces its adherents to define themselves in relation to others, rather than by their own peculiar combination of ideas; and the others have to be invented, even caricatured for the purpose. The point about an open world is that there are not just two or three ways. There are ... 101 ways, which is to say, an indefinite number.
> (Dahrendorf 1999)

Clearly, despite Giddens' protestations of openness, the third way implies that it understands the next phase of human development and thus can and should control that development. Thus, if the world has changed (due to globalisation, reflexivity, etc.) and the third way is capable of understanding that shift, then the third way can become the new 'radical' historical order which saves the Left and restores it to its position at the forefront of history.

Given this combination of policy openness and underlying order and control, one can easily see why the third way would be so appealing to centre-left parties like Labour and the Democrats, and even some more traditional continental European parties. In the UK it allowed 'new' Labour to justify its abandoning of unpopular traditional strategies (dropping Clause 4 and accepting the Thatcherite marketisation of much of the UK public sector) under the veneer of a vaguely defined new order. In policy areas, it legitimated the ability of the Blair government to pursue seemingly decentralising policies in a context of centralised audit and control. The most blaring examples occurring with the welfare state, particularly in the education and health sectors, where the calls for local control and autonomy have been strangled by the centralising flurry of continued 'new managerialism' (Walsh 1995) and an intensified audit culture (Walsh 1995; Rouse and Smith 1999).

How does complexity differ from and go beyond the third way?

For Giddens, as human actions have increasingly come to dominate the natural world, the interface between humans and nature has become increasingly complex. This new manufactured uncertainty is further complicated by the increasing social reflexivity of individuals in the post-modern world. Not only do they have to confront manufactured uncertainty, but they are no longer willing to believe in or submit to traditional authorities or ideologies. Hence, traditional ideologies of Left and Right are increasingly outdated and useless in the current age.

On all of these aspects, complexity thinking would agree, but go a step further. Yes, traditional ideologies are outdated, manufactured risks have increased and individuals have become more socially reflexive. However, complexity would stress that the natural world and external risks were never completely orderly. In fact they have always been complex interactions between humans and nature with unpredictable consequences. Human interaction with the weather, plants and animals, to take just three obvious examples, has led to a multitude of complex outcomes throughout the history of humanity, from the plague to the potato famine. Consequently, there has always been a degree of manufactured risk and human beings have attempted to deal with it in a similar ways. Even in pre-modern societies, human beings attempted to deal with manufactured risks by promoting strategies of social order. Countertendencies to these strategies emerged and even reconciling 'third ways' (Smith 1973: 134). Thus, when critics of the third way complain that it is nothing new, they are more correct than they know.

Politically, what is distinctive about the Newtonian framework of the eighteenth, nineteenth and twentieth centuries was the degree to which humans believed that they could order their societies. Flushed with the heady success of mechanistic and industrial achievement no problem seemed beyond the grasp of humanity. Social scientists merely wedded this vision to the social realm and produced the fundamental visions of social order, communism and capitalism, which structured the history of the twentieth century. From a complexity perspective, both visions of global order were never possible. Pure communism where the state dominated every economic transaction and pure capitalism where the market determined all social interactions were equally unsustainable within the complex interaction between humans and nature.

At first glance, Giddens' third way seems to pursue a similar strategy regarding Left and Right. He is critical of both market and state extremes and produces a raft of policy proposals which combine elements of both. However, he also wants to recapture, for the Left, its earlier 'future-oriented radicalism' (Giddens: 1994, 73). Here we see the fundamental contradiction at the heart of the third way. On the one hand, Giddens wants to break with the determinism of the Left which argued that capitalism led to socialism, proletarians were humanity's saviours and history had a clear direction towards some fixed final order (called socialism/communism). A complexity perspective would certainly agree with that. On the other hand, his desire to find a radical 'new way' that gives the Left back its position at the forefront of historical development has clear overtones of earlier twentieth century attempts to create a predetermined final order. Hence, by not recognising the full implications of complexity he opens up the third way to criticisms that it is both amorphous and authoritarian at the same time.

From a complexity perspective, beyond creating a stable fundamental order within which individuals can learn, interact and adapt, there is little a state can do with complete certainty. Most policy arenas can be understood as evolving complex systems that display fundamental regularities, but their particular dynamics cannot be known with certainty. Detailed study and knowledge can reveal more, but cannot create pure orderly knowledge. A society that is stable, open, democratic and encourages complex interaction is likely to be much more successful than a closed strictly ordered society, or a destabilised chaotic one. However, complexity cannot predict which type of similar societal organisation (the more market-oriented British, corporatistic Germans, socialistic Scandinavians, etc.) will be more successful than another. From a complexity framework, there are no certain strategies other than the most fundamental ones and as Dahrendorf stressed there are not three ways, but a hundred.

Similarly, since there are so many possible 'ways', there is no particular reason why the US/UK, or any other state, should be seen as the leader of the third way. A particular set of strategies may work in one case, but not in another. One can make moral arguments over which system one may prefer, Scandinavian or British, but no state can claim to have the one and only 'way'. More fundamentally, there is no reason for the third way to be an inherently leftist strategy. A complexity framework implies uncertainty for both Left and Right. The two political movements have distinctive values that imply particular policy strategies. Nevertheless, they are both caught within the emerging complexity paradigm. Giddens' desire to bring the Left back to the forefront of history, betrays his inability to break with the earlier Newtonian framework and is surprisingly similar to the 'end of history' concepts of the right-wing Francis Fukuyama. It may be politically appealing to the Left, but is theoretically and practically unsustainable. The future is not a search for a third way, but for any number of 'third ways'. It is not a politics of the search for a new order or disorder, but a politics of continual complex choice.

Complexity, politics, and the third sector

Keeping with the theme of 'thirds', another area where the implications of complexity can be easily explored is in the general debates over the role and impact of the 'third sector'. The third sector (sometimes called the voluntary sector) is a somewhat vague, but useful term for labelling the mishmash of organisations and social groups that sit between a profit-oriented market sector and a bureaucratically driven state sector (Evers and Laville 2004). Third-sector organisations range from charities, trusts and foundations to mutuals, voluntary organisations and member and producer co-operatives. They also represent a diversity of internal organisational forms, ranging from large hierarchical structures to small-scale local organisations. Moreover, the size and structure of particular third sectors varies substantially between nation-states. Spain, Italy, Portugal have powerful co-operative movements (worker, members and producer) while others, particularly Germany, influential third-sector charities based on churches, political parties and trades unions. Meanwhile, the British and Swedes have relatively limited third-sector movements, while the French have a complex system of well-defined national, regional and local organisations. Some elements of the third sector have deep historical roots, stretching back to the mid-nineteenth century. However, other parts have only emerged in last 20–30 years (Osborne 2008).

The well-documented growth of the European third sector in recent decades is intimately linked to fundamental European political and economic transformations including: economic internationalisation, the diversification of post-industrial societies, the failure of statist strategies to create acceptable socio-economic outcomes and the inability of the market to function without appropriate state intervention (Kendall 2003; Pierson 1994). Like the 'third way', the third sector seemed to be a way for societies to go beyond the statist-market dualism and openly combine different strategies to obtain acceptable outcomes. The growth of this sector was generally encouraged by third way thinkers who saw it as an added arena for the state to fulfil welfare criteria and social needs.

But this third way interpretation does not go far enough. From a complexity perspective one needs to treat the third sector like a complex system if it is going to survive and prosper and recognise that the third sector is an essential element in the 'symbiotic competition' between the state and market which enables all three systems to successfully co-evolve and adapt to the ever-changing global and European political economy. It is also an essential strategy for allowing stakeholders to participate more actively in decision-making.

Seen as a complex system, the third sector is an example of a robust and creative social system. Its organisational diversity and adaptability is a clear indication of its fundamental strength and creative potential. As any complex system, in order to thrive it must find the zone of creative complexity between stultifying order and destructive disorder. In order to do this, it needs a fundamental framework within which to operate and the individual units need to have the ability to interact and learn so that they can survive and adapt. Basically, there is no perfect organisational form for the sub-units in the third sector. They adapt to their given surroundings. The more they are able to learn and adapt to their fitness landscape, the more likely they are to prosper. Thus, prescribing a given form to all third-sector organisations would virtually guarantee the systems failure, robbing it of its ability to adapt and learn. Likewise, limiting the forms of interaction between third-sector actors and other actors would also constrain their adaptive capabilities. On the other hand, the third sector will never prosper if it is caught within an unstable framework. Constant large-scale shifts in the basic legal/regulatory framework would limit the learning capabilities of third-sector actors and punish the essential risk-taking strategies that are necessary for organisational learning and innovation.

In essence, the third sector already acts as a healthy complex system engaged in creating evolutionary learning and adaptation resulting in successful creative complexity. Not only does the third sector function as a complex system, but through a process of 'symbiotic competition' it actively increases the complexity of the state and the market. Using a complexity frame of reference, if one views nation-states as complex organisms composed of three main interacting socio-economic systems (market, state and third sectors), then the most successful evolutionary strategy would not be to pick one system and promote it above the others, as statist or free market supporters have attempted throughout the twentieth century. These reductionist strategies, based on orderly assumptions and convinced of the inherently superior nature of one form of societal organisation over another, automatically reduce the complexity and adaptability of their society leading to 'stultifying order'. Whether from a statist position, demonstrated by the failure of French statist economic strategies of the early 1980s, or a market position, confirmed by the high social and political costs of the Thatcherite strategy in the UK, simplistic approaches to complex evolutionary societal dynamics are bound to lead to negative outcomes.

From a complexity perspective, the most successful strategy would be to create a 'symbiotic competition' between the systems. No system would be inherently preferred above another. All would be encouraged to compete over the creation and distribution of socio-economic goods. Within this general framework, local organisations would be encouraged to interact, learn and evolve with a minimum of centralised interference. Historically, as mentioned above, attempts to radically order society along market or statist lines have tended to automatically reduce societal complexity and have proven to be unsustainable in the long term. The advanced industrial economies have prospered in the nineteenth and twentieth centuries within a very broad framework of open societies and market economics. Nevertheless, the multiple paths that led to this common success were extremely varied, ranging from the more market-oriented Anglo-American model, to the corporatist Northern European, to the family-based Asian model. In complexity terms these results are remarkably similar to the healthy variety demonstrated by adaptive organisms pursuing one stable *attractor*. Central to this success has been the ability of each nation to pursue their individual evolutionary strategies within a stable, but non-stultifying framework. Certainly, during the twentieth century, the third sector has played a key role in this development. By offering an organisational alternative to market and statist systems, it forces them to adapt, learn and evolve. At the same time, the third sector adapts and learns to compete with the other sectors, hence symbiotic competition. Where one or more of these systems have been severely repressed, average complexity has been lost and socio-economic evolution has tended to stall.

In sum, for many academics and policy practitioners, the third sector is the ugly step-sister of the statist and market systems, tolerated but resisted. It was often seen as an indication of state and market failure. From a complexity frame of reference, it is not only an essential element of advanced industrial economies, but significantly enhances the learning, adapting and evolving capabilities of the market and state systems.

Is the EU compatible with the third sector?

Linked to this issue is the interesting question of whether or not the EU, that was originally designed as a predominately market creating institution, is compatible with the third sector. In general, the EU is an excellent mechanism for enhancing the complexity of EU member-states and Europe. Obviously, the EU did not emerge out of an attempt to promote European complexity. Inter-state bargains, sectional interests and federalist idealism all played a role. In European integration theory, debates over the fundamental dynamic of integration followed typical linear models of the 1950s and 1960s. These theories fell in and out of favour from the 1950s to early 1980s (Rosamond 2000). However, with the revival of the EU in the mid-1980s theorists began to reassess the meaning of integration. They began to recognise that the EU had several fundamental dynamics working at the same time (Sandholtz 1996). Building on this, theorists began to turn towards more flexible concepts of governance and multi-level governance arguing that the EU was a multi-level playing field which exhibited different organisational outcomes and policy dynamics depending on the particular policy area.[5]

Unsurprisingly, this theoretical evolution reflected the policy transformations that were taking place within the EU. During the 1980s and 1990s, as the EU was experiencing unprecedented organisational growth, it became increasingly obvious that the earlier rigid policy approach of harmonisation and convergence was unsustainable. More flexible concepts like 'mutual recognition', 'subsidiarity' and 'best practice' began to emerge. In the 1990s, new 'soft' legislative and legal instruments were increasingly used (Cram 1997), particularly in the areas of social and employment policy (epitomised by the Employment Guidelines and Open Method of Coordination). In addition, the EU began to try and promote a European level social and civil dialogue with emerging European wide social actors (Geyer 2000). In essence, rather than attempt to force rigid forms of socio-economic organisation onto the member states, the EU evolved an appropriate complexity strategy of providing a general framework for interaction between the different systems.

It is important to note that this was not a planned result, but a combination of the weakness of the EU institutions and the desire of the member-states to create an orderly framework for market interaction. Had the EU attempted to centrally impose a rigid organisational form on the member states, the member states would have rejected it and the system would have broken down. Fundamentally, the very messiness and complexity of the EU gives it the flexibility to evolve and adapt and to promote symbiotic competition between different member-states' market and state systems.

As the EU began to evolve in a more complexity-oriented direction, it began to recognise the importance of the third sector. The EU began to integrate the third sector into its proposals for the promotion of employment, social inclusion and anti-poverty initiatives (Geyer 2001). Moreover, elements of the third sector were increasingly integrated into the political process through the civil dialogue and the growing consultative nature of the EU political process. Fundamentally, as the

EU evolved it began to look beyond the limitations of market and state system integration and explore the potential of third-sector integration. Thus, the growing complexity of the EU should encourage it to explore and support third-sector development.

Using The 'balance and range of outcomes' tool to help the EU promote the third sector

As discussed in Chapter 3, the primary strategy for complex systems is to stay within a zone of creative complexity as much as possible and avoid destructive disorder and stifling order. From a complexity standpoint, the EU is already doing a number of things to promote the third sector. By being open to third-sector alternatives and utilising them in the policy process, it is already promoting their development (Kendall and Anheier 2001). Nevertheless, a number of fundamental strategies must be continually kept in mind.

- *First, do not overly order the third sector.* As noted above, the third sector varies significantly between nation-states. Moreover, within each national system there are a significant variety of organisational forms within the third sector. This variety is an indication of the healthy complexity of the third sector. The EU should avoid trying to pick or promote a given type of third sector or organisational form for third-sector actors. Fundamentally, the EU will have no way of knowing whether a given form is correct for all situations. The attempt to impose such a form would lead to a political backlash. And, if successfully imposed, it would significantly weaken the complexity of the system.
- *Second, promote symbiotic competition between national level third sectors.* Most third sectors in the member states are dynamic learning systems within structured environments. However, if these systems are to increase their complexity and hence learning, adaptation and innovation skills, encouraging interaction between the national systems is an obvious step. Again, this should not be imposed from above through a strategy of harmonisation, but through an opening up of the third-sector activities between member-states. Examples could include the ability of third-sector organisations to organise and/or offer services in other member-states (social care and young and elderly services are obvious possibilities). In essence, this implies a 'mutual recognition' of the third sectors in the various member states and the rights of individuals and societies to access and take advantage of different systems.
- *Third, promote the third sector at the European level.* Obviously, not only should the EU promote interaction between the national third sectors, but it should also promote the creation of a European third sector. In many ways, one could see this as a similar project to the creation of a European free market. National systems would continue, but would be encouraged to learn and innovate through interaction with other national systems and a European system. The European third sector could be promoted through regulatory and

legislative proposals and through basic funding. An example of this type of support already exists in the EU's promotion of the social and civil dialogue and funding of social and civil European actors.
- *Fourth, promote the third sector against the market and state.* Of the three main sectors, the third sector is the weakest at the European level. Market interaction and integration has accelerated radically since the 1970s. Similarly, state interaction within the EU institutions and between the member states has increased dramatically since the 1980s. On the other hand, the European-level third sector is in its emerging stages. There is little intra-member-state system interaction and uneven development at the European level. If the EU were to significantly promote a European third sector, European market and state actors would be likely to feel threatened and oppose the development. In response to this, the EU should do what it can to protect the third sector from these fundamental threats.
- *Fifth, keep the focus on interaction, adaptation, learning and innovation.* A final key point to remember is that, from a complexity perspective, there is nothing inherently supreme about the third sector. The sector provides a number of socio-economic benefits and offers a number of distinctive strategies for social learning and development. However, complexity theory would argue that its primary use is as a strategy for promoting interaction, adaptation, learning and innovation within the other main social systems and within itself as well as promoting stakeholder involvement at all levels. Complexity is not wedded to any particular type of social organisation. What it does say is that the society that is the most complex, with the largest amount of interaction, adaptation, learning and innovation going on within it and can stay within the zone of creative complexity for the longest will be the most likely to succeed. Consequently, the third sector is not an end by itself, but a strategy for promoting the complexity of European society.

In sum, with complexity theory one has a 'scientific' foundation for the importance of the third sector. Not only is the sector important, but through symbiotic competition it is vital to the market and state systems as well. The EU, due to its multi-level and decentralised structure, tends to pursue complexity maximising strategies that often involve the third sector. Thus, there is a general institutional affinity between the EU and the third sector. Moreover, from a complexity perspective, EU policymakers should actively promote the third sector at the national, intra-national and European levels and encourage its interaction with the market and state systems in order to maximise the complexity of the European economy and society.

Conclusion

Hopefully, after looking at political topics as diverse as democracy, the third way and third sector, it is becoming increasingly clear that complexity has vast implications throughout the entire range of politics and so-called political science.

In essence, what we are arguing is not that politics should give up its attempts to be scientific, but recognise that 'science' has not stood still. Happily, it continues to emerge and evolve and that learning from recent developments in complexity can be a way to significantly reinvigorate the study of politics. This does open us up to the criticism that we are merely duplicating the 'science envy' that got politics into the Newtonian orderly trap that it is stuck in now, merely using a new rather than old type of science. However, we would argue that the openness and uncertainty inherent in the basic concepts of complexity provides a more realistic framework for understanding the human world and that, eventually, yes, we would expect a new post-complexity framework to emerge in the future. What that will be we have no idea. What complexity can accept is that there will be something different.

5 Health

> To cope with escalating complexity in health care we must abandon linear models, accept unpredictability, respect (and utilise) autonomy and creativity and respond flexibly to emerging patterns and opportunities.
> (Plesk and Greenhalgh, *British Medical Journal* 2001)

Following on from our discussion of politics, we now turn to one of the most complex policy areas: health. Despite its obvious 'commonsensical' complexity, health policy and healthcare have not escaped the relentless impact of the orderly paradigm. From remote policy-makers and local hospital managers to patients and carers, a belief in the mechanistic nature of health exerts a distorting influence. Policy actors believe that more money will create a better health system and lead to improved health. Doctors are adamant that patients should do as they are told (so-called 'patient compliance'). Meanwhile, patients are convinced that their health problems are fundamentally 'mechanical' and that an experienced doctor, probably located in a high-tech hospital with state-of-the-art equipment, should be able to 'fix' them – just like a car.

Adoption of a complexity framework, we believe, would help to move beyond this static understanding. In this chapter we will look at three levels of the health spectrum: global policy, national health systems and individual chronic disease management and show how complexity helps one to rethink the fundamental approaches to all of these areas and provides health decision-makers and individuals with new tools to understand how to move beyond the limits of traditional orderly concepts of health.

The influence of the paradigm of order on health

As discussed earlier, the foundations of the orderly paradigm were quickly integrated into all fields of the human sciences. Health was no exception. From an orderly perspective, the body became just another type of mechanism – much more complicated than clocks, but fundamentally no different. The trick to understanding and controlling the body (correcting its mistakes) was to reduce it to its component parts and find tools for understanding and measuring its 'motions'. Whereas in the seventeenth century, medical diagnosis was based heavily on listening to the patients interpretation of his/her condition, by the eighteenth and nineteenth centuries a growing array of technological innovations, such as the microscope, stethoscope, blood pressure monitors, x-ray machine, and so on, 'encouraged a physical separation of the doctor from his patient' (Reiser 1978: 90). With the growing technological advances of the late nineteenth and early twentieth centuries, medical laboratories, centralised hospitals and the routine use of a wide array of medical testing procedures began to proliferate. Testing became not only a way of objectively evaluating the patient and protecting the doctor from charges of malpractice, it even allowed doctors to feel that they were 'using the same rigorous methods as did the scientist who pursued truth in his laboratory' (Reiser 1978: 162).

Linked to this 'scientification' of medicine was the growth of specialisation. This was the logical outcome of the growing detailed knowledge of the human body combined with the belief that it was fundamentally a mechanical clock that could be separated into its key components. Large-scale medical institutions would bring together all of these specialisms into a 'one-stop' body shop! Eventually, with the growth of computers, more and easier testing and greater knowledge, disease could be eliminated as it was on the science fiction TV and films of the 1960s and 1970s.

In health policy, this translated into a belief that the best health systems were the ones that focused primarily on the most technologically advanced aspects of health and what large-scale, highly specialised health institutions (hospitals, research centres, etc.) could do for health (the area of 'tertiary care'). Unsurprisingly, assuming that what worked in the West should work for everyone else, during the 1950s, 1960s and 1970s developing countries were strongly encouraged to mimic this high-cost, technological strategy (Kiely and Marfleet 1998). Provision of large-scale systems of advanced tertiary care became not only a policy for improving health outcomes but a symbol of the country's progress towards a fully developed state.

In the wake of the 1970s and 1980s debt crisis in much of the developing world, the World Bank (supported by USAID) began to play a fundamental role in the definition of global and national health policies. In its influential 1993 *World Development Report* it laid out a unified approach for national health policies around the creation of large-scale privatised health systems, arguing that national governments should 'limit government action to formulating policies, providing a limited package of public health interventions ... developing and enforcing regulations and financing essential clinical services for target groups' (Brugha and Zwi 2002: 65). Once again, Western experts building on the logic of the orderly paradigm concluded that the root of global health problems was that national health systems were not similar enough to the standard neo-liberal model. Even though in reality the various health systems in the advanced economies were neither as unified nor market-oriented as the model advocated. All of this is not to imply that there have been no voices challenging the dominance of the orderly framework in health. As early as 1876, Theodor Billroth was worried that:

> the object of our modern endeavor is to make the physician's skill ... independent of the talent of the individual and may be reduced to an absolute science. All knowledge and skill are to be determined and controlled by means of the laws of arithmetic and logic in order to make everything absolute and permanent ... I doubt if this good will ever be reached, at least in the art of healing.
>
> (Gladston 1981: 105)

More generally, it has been recognised for quite a long time that health is closely affected by nutrition, clean water supplies, and good sanitation. Even when considered at this superficial level, it is reasonably obvious that the health field with its diverse layers of interconnected elements could not possibly be served adequately by one setup founded mainly on a series of hospitals geared to receiving a flood of sick people referred by general practitioners. The other factor in the equation; keeping people healthy for as long as possible, is clearly a highly significant element. And keeping people healthy is not a simple matter. As already mentioned, a number of factors acting alone and in combination determine a person's state of health. Their effects are not instantaneous, some stretching into decades. The factors themselves are not static. New discoveries are being made all the time, from the influences of food additives and new products, to environmental threats and changes in lifestyles and fashions. Without going deeply into the science of complexity, and using the word in its common parlance, the health field does seem to be unquestionably complex and simple, orderly remedies such as more money, further reorganisation, more use of information technology, greater involvement by the private sector and imposed performance targets have all been tried and found wanting.

Better health at a fraction of the cost – focusing on the basics

> Many deaths of children under 5 years of age could be averted for $10 or less ... but the average actual expenditure in poor countries per death prevented ... is $50,000 or more.
>
> (World Health Organisation's *World Health Report* 2000: 10)

This statement is remarkable only because it mirrors conclusions from numerous other reports and textbooks. Money does not buy effectiveness. In health terms the magic mix is promotion of public health, focus on primary care, and collaboration between providers of local services such as health, housing, employment and social agencies.

Taking just one famous local example, the city of Liverpool's history suggests that the basic facts about the complexity of health have been known for a long time. The health of citizens was in a dreadful state at the height of Britain's capitalist-driven Industrial Revolution. Two men appeared on the scene that had more effect on health than all the hospitals and doctors put together. The first was William Henry Duncan (1805–63); appointed Medical Officer of Health in 1847, and James Newlands (1813–71); appointed as Britain's first Borough Engineer also in 1847. They worked closely together on an agenda that recognised the key links between housing, water supply, sanitation and health. Duncan helped to draft Liverpool's first Sanitary Act and Newlands constructed the world's first integrated sewerage system. The end result? Average life expectancy doubled during their lifetimes; from an incredible low of 19 years![1]

Interestingly, the same line of thought was in evidence across the Atlantic at almost exactly the same time. The Report of the Sanitary Commission of Massachusetts was presented to the state legislature in 1850. It recommended:

> immunisation and communicable-disease control, promotion of child health, improved housing for the poor, environmental sanitation; training of community-oriented health manpower; public-health education; promotion of individual responsibility for one's own health; mobilisation of community participation through sanitary associations; and creation of multidisciplinary boards of health to assess health needs and to plan programs in response to sound epidemiological evidence.
>
> (Evans *et al.* 1981)

In effect, a battery of measures focused on encouraging, and maintaining, a desirable direction of travel: an inherently complex view of health.

Efforts continued to try and wean health authorities off expensive secondary and tertiary care that have undoubted but limited impact on the health of communities and to focus their attention on other more effective interventions that relied on public health, primary care, and provision of safe water, sanitation, housing and adequate nutrition. In 1978, WHO and UNICEF held an international conference on the subject in Alma-Ata (now Almaty in Kazakhstan). All the pros and cons were rehearsed but ultimately the impact was limited (WHO's World Health Report 2000: 14). Basically, the professionals knew best: they decided what people needed, rather than what they wanted, and set out to satisfy the presumed needs through high-cost, but less effective, means. Decades have been wasted in pursuing inefficient policies in the name of 'science'.

An efficient national health system: the Cuban case

Too much money could be just as bad as too little. For as long as they could afford it 'wealthy' countries stuck to an outdated and ineffective perception of what health was about. Pressure on budgets forced them, albeit reluctantly in some cases, to change direction.

The same transformation took place in Cuba decades ago. In that case, harsh sanctions imposed by the US on the island in the wake of the communist revolution led by Fidel Castro brought the economy to its knees. Extravagant health spending was not a viable option. Moreover, in order to maintain the popular base of the political regime, allowing the health of Cubans to deteriorate was not acceptable either. Imaginative policies and practices had to be adopted. It goes without saying that old ideas, not unlike those advocated by Duncan and Newlands in Liverpool and Shattuck in Massachusetts, were picked up, dusted off and then put in practice. The results were spectacular. Today, as confirmed by a plethora of WHO and UN reports and seen in Michael Moore's film, *Sicko*, Cuba has a first-class health system where life expectancy is now higher than the US and per-capita health expenditure is less than one-twentieth of the US level! A strong hint of the foundation for this success lies in the fact that Cuba has had a Ministry of Public Health since 1961.

One of the most detailed studies of Cuba's social services – including education, health and sanitation – was completed in January 2002. It was commissioned by the World Bank as background for the 2004 World Development Report. The report underlined the fact that 'monitoring the health data of the population plays an important role in the evaluation and shaping of health policy'. Comparison with the dearth of reliable information within the British NHS is instructive. A comprehensive picture of Cuba's state of health is included in the form of appendices to the report.[2] Basically, life expectancy increased from 64 in 1960 to 76 in 2001. Similar positive views were expressed in summer 2002 in the *Harvard Public Health Review*.[3]

Significantly, Cuba's health system evolved through a number of stages over a period of five decades; municipal polyclinics, the 'medicine in the community' programme, and the 'family doctor-and-nurse teams'. This reiterative process is typical when dealing with complex adaptive situations. Clearly, the Cuban case illustrates that effectiveness does not necessarily depend on funding. Much could be achieved at low cost once the right policies have been defined, implemented, monitored and then modified sensibly.

Is Cuba's experience unique and what is the magic formula?

Far from it: there are other instances that present similar, or possibly even more impressive, outcomes derived from adoption of practices that are radically different from those based on the traditional health and healthcare viewpoint founded on an orderly linear formulation. Basically, the cost of achieving good health (and education, sanitation, nutrition, ...) could be so low it would be, to a degree, irrelevant to decision-making. However, this requires a major change in approach.

The above seemingly surprising conclusion is not a new discovery. Evans *et al.* reported in 1981 that China, Sri Lanka, and the state of Kerala in India 'are examples of countries that have attained a life expectancy close to the level in the industrialised world, with income levels in the range of the least developed countries' (Evans *et al.* 1981). Similarly, Lloyd Timberlake wrote, 'African countries can expect the greatest improvement in life expectancy from health investments in materials and child health services in rural and urban slum areas, costing less than $2 per capita' (Timberlake 1991: 40).

The solution to the riddle is simple and self-evident. Fundamentally, the 'Horses of the Apocalypse' hunt together. The basic problems faced by most people on earth are closely linked. Reducing one problem could also help to ameliorate others. On the other hand addressing several problems at the same time through modest measures would yield unexpectedly impressive results. The whole is greater than the sum of the parts: a highly significant feature that indelibly marks the way complex adaptive systems work. This is particularly noticeable in the case of health, nutrition, water quality and sanitation.

> Around 7 out of every 10 deaths among children under the age of five in developing countries can be attributed to a few main causes: acute respiratory infections, diarrhoea, measles or malaria. Malnutrition contributes to about half of these deaths ... A child deficient in vitamin A, for example, faces a 25 per cent greater risk of dying.
> (UNICEF, *The State of the World's Children* 2005)

What is the cost of vitamin A? The answer has been known for long:

> Daily diet can be changed, usually at little cost, to include small amounts of green leafy vegetables; or 2 cent vitamin A capsules can be given three times a year to children over six months of age, or vitamin A can be added to sugar or cooking oil.
> (UNICEF, *The State of the World's Children* 1995)

Lack of iodine is just as damaging to health; including cretinism and mental impairment, and equally cheap to remedy. Again, the easy way to deal with this problem has been known since the turn of the twentieth century: minor change to diet or iodine added to salt at a cost of 'about 5 cents per person per year', UNICEF estimated back in 1995.

In general, we have known for decades that poorer nations require about 50 billion dollars a year to get out of the mess that goes for living in these countries. We also know that whilst it is easy to find the cash to mount wars – the recent war on Iraq has cost over one trillion dollars already – it is almost impossible to find the money to help poorer nations. In the meantime, cheap measures that would produce near-miraculous effects on the standard of living of four out of every five people on earth are obstinately shunned in favour of large-scale solutions based on 'modern' models of health seen in advanced industrial societies.

As mentioned earlier, there is nothing new about this message. Complexity merely provides a theoretical framework for understanding why these simple local actions can be so important and effective. Complexity, by itself, will certainly not change the minds of corrupt elites. However, it does help to scrape away the 'modern' and 'scientific' veneer that they often cloak their self-interested policies in. Hopefully, with a growth of complexity, more sensible policies and actions

98 *Health*

may have a chance. The measures are certainly not dramatic and they will not grab the headlines, but is this a good-enough reason to condemn the majority of the world's population to a life of abject misery?

The case of the English National Health Service[4]

Now, how does complexity apply to wealthy national health system and what does it imply? In terms of large-scale public policy, the English National Health Service[5] has been a successful complex system for much of the post-WWII period. Established by a reforming Labour government in 1948, the NHS 'nationalised' the diverse provision of health care into a unified tripartite structure based on community health services, primary care (local doctors) and hospitals. The system was similar to other 'nationalised' health systems that emerged in Europe, produced reasonable results in terms of life expectancy, health care and health outcomes and up until the 1970s went through a sensible range of 'reforms' and 'reorganisations'. However, as we will briefly detail below, this process was radically accelerated from the 1980s onwards resulting in over 20 major reforms and reorganisation (Walshe 2003), leading one top health expert to conclude that each NHS reorganisation had 'a half life of two to three years before it is abolished or displaced by another' (Glasby *et al.* 2007: 3).

This process of 'redisorganisation' (Oxman *et al.* 2005), driven by a traditional centralised hard managerial approach, belief in end-state planning and continual fluctuation of political and managerial elites has significantly destabilised the NHS, lowered staff moral, increased management costs, constrained local autonomy and weakened its 'public spirited' organisational culture. In essence, by refusing to recognise its complex nature, British central policy actors are busily crippling one of its most successful policy innovations.

A bit of NHS history

As detailed by Gorsky (2008), the history of the NHS can be divided into three phases: 'foundation' (1948–79), 'Thatcherisation' (1979–97) and 'New Labour' (1997-present).

During the first phase there were a number of minor changes to the basic structure; such as funding fluctuations in 1950s and 1960s, and increasing emphasis on primary care with a practitioner contract concluded in the late-1960s. This period, however, saw one large project; the 1962 Hospital Plan, and one major reform; the 1974 NHS Reorganisation. The aim of the 1962 Plan was to create a hierarchy of specialised teaching hospitals and more local district general hospitals (DGH) and subsidiary health centres. Enoch Powell, the Minister of Health, boasted at the time that the government was planning the hospital service on a 'scale not possible this side of the Iron Curtain'.[6] Financial and operational pressures resulted in scaling down of the project and affected its ambitious timescale, but the main components of the plan are still visible to this day. The 1974 NHS reorganisation created 'area health authorities', with mostly the same boundaries as local authorities to bring health care and social care closer together.

During the Thatcherite phase, with the rise of neo-liberalism and market-oriented norms the NHS was seen as particularly ripe for change and reform. Political elites focused on obsessive bouts of reorganisation aimed at treating the health care system as a business. The new format relied heavily on powerful management from the centre. Consensus was seen as an obstacle to fast implementation of difficult 'reforms'. The package of changes, moreover, sought to separate purchasing (commissioning) from provision of health services coupled with early attempts at attracting private sector providers and creating an 'internal market'.

The 1983 Griffiths management review, *NHS Management Inquiry Report*, was a key event in this phase of NHS history. As Griffiths admitted, 'We had not been asked to prepare a report, but that we should go straight for recommendations on management action ... Speed of implementation is essential.'[7] Moreover, the authors of the review made a telling observation: 'In short if Florence Nightingale were carrying her lamp through the corridors of the NHS today she would almost certainly be searching for the people in charge.'[8] Assertive management linked to fundamental belief in clear criteria and direction lay at the heart of reforms in this period.

The current 'New Labour' phase saw a substantial increase in overall funding for the NHS coupled with a plethora of central targets and controls. *Shifting the Balance of Power*, published in July 2001 went back to the not-unexpected idea that primary care should play a bigger role. That was followed shortly thereafter by the creation of 303 Primary Care Trusts (PCTs) 'to improve administration and delivery of healthcare at local level'.[9] Inevitably, a 2004 NHS Improvement Plan focused on 'putting people at the heart of public services' and, later, Independent Sector Treatment Centres (ISTCs) were introduced to encourage further private sector presence under the NHS umbrella.[10]

Another shake-up reduced the 303 PCTs to 152 in 2006; Strategic Health Authorities having been reduced from 28 to 10 earlier in the same year. The boundaries of the new PCTs were almost identical with those of the local authorities bringing the boundary situation back to 1974! At the time of writing, PCTs are undergoing yet another major change: spinning off their provider functions to complete the separation between purchasing (commissioning) and provision.

More fundamentally, during this period centrally set targets were rigidly monitored by the Healthcare Commission. This audit/target culture attracted widespread criticism. Partly in response, a document entitled *Standards for Better Health* document[11] was issued in 2004 with the declared aim of cutting targets. These were renamed standards and were reduced to 'only' 7 domains; with 24 core standards and 37 sub-headings, coupled with 13 developmental standards; with 21 sub-headings!

Following the appointment of Lord Darzi as Health minister, a major reorganisation of health services in London was developed and then extended to the rest of England. The *Next Stage Review* (2008)[12] and subsequent *Transforming Community Services* (2009)[13] advanced ideas that no one could disagree with: quality care, patient-led service, clinician involvement, NHS constitution, vision, training academy, and so on. Taken individually its many ideas are self-evidently sensible, but they are conceived as interdependent components of one central plan; the ultimate in clockwork reductionism. The components must come together at the same time and in the correct manner for the overall plan to work. Hospitals are redesigned, and reduced in size, to fulfil precise functions. Primary and community care are reconfigured to do much of the work previously undertaken in hospitals in addition to their traditional activities. Functionally appropriate buildings for the new plan must be ready on time to welcome the influx of new patients and, even more of a challenge, clinical and administrative staff in the exact skills and numbers must be there to treat them. There is no margin for error and no going back.

And the result

Despite being increasingly recognised in the academic literature[14] and even in the NHS's own internal publications[15] as a complex system, the impact of overly centralised hard management styles, belief in end-state planning, fluctuating political manipulation and the audit/targeting culture continue to afflict the NHS. These orderly constraints are impossible to reconcile with the proper functioning of such a large, complex institution. An end-state plan by its very nature is a fixed entity that has to be implemented in a way that allows the parts to fit perfectly in the right place and at the right time, clockwork fashion. A change in the governing party or personalities cuts across that assembly-line approach. Similarly, centrally defined end-state plans have to gain acceptance at all levels in a human activity such as healthcare in which there is often close personal involvement by both patients and clinicians. Furthermore, convincing staff at all locations to fall in line with a continual stream of what seemed like constantly changing demands for reform bombarded at them from the centre has not been possible. In effect, there has been a perpetual and unresolved tussle between those at the top and those on the front line.

These conditions revealed two features which are self-evident to those familiar with complex systems but caused much consternation within the NHS. On the one hand, during the past two to three decades the NHS's 'organisational culture' exerted itself forcefully in an effort to maintain the integrity of the service traditionally provided by the NHS in the face of the disruptive agitation for endless so-called reforms from the centre. As Jake Chapman pointed out, under external pressure for change, an organisation 'will remain recognisably the same institution. What is conserved is its internal organisation, core values, and culture' (Chapman 2002: 41–42). In the case of the NHS this was interpreted as bloody minded resistance to reform resulting in regular calls by minsters and their advisors for a 'culture change'. Griffiths, tasked with a 1983 'reform' commented: 'To the outsider, it appears that when change of any kind is required, the NHS is so structured as to resemble a "mobile" designed to move with any breath of air, but which in fact never changes its position and gives no clear indication of direction.'

The other feature revealed by the flurry of reforms was yet another concept familiar to those working in complex systems. The different ideas, seen by their originators as radical attempts at reform cycled, in the main, through one *attractor* that enveloped the different states that the NHS could assume at any point in time.[16]

It is essential to point out here that the NHS, a complex system subject to internal and external influences, has to and does evolve. However, reforms based on central planning founded on command-and-control from the top by political groups and individuals governed by electoral timetables and periods of tenure does not allow for experimentation, innovation, learning from experience, constant fine-tuning of problems and solutions and, if necessary, a change in direction of travel. Often, political and policy elites see these contradictions but instead of taking the next step and engaging fully with complexity they try desperately to stick to the old failed remedies.

For example, the Darzi reorganisation tried to combine continued centralised control, targeting and end-state planning with increasingly diversified provision. This included separating purchasing (commissioning) from provisioning of services and then required that provision at all three levels; primary, community and hospital care be sought from a wide range of public and private organisations. The transition from an integrated NHS into the 'new' NHS was to be achieved seamlessly in a few years. These potential risks have accumulated as each individual idea was bolted onto the wider system without much thought to system-wide consequences triggered by interactions between the multiplicity of components old and new.

A clear example of the risks of such an approach can be found in the substantial management information setup needed for a dispersed model of commissioning and provision to operate effectively and efficiently. A National Programme for IT (NPfIT) was launched by the NHS in 2002 to hold, process, and maintain records on some 60 million people. The original cost was set at £2.5 billion; a modest sum for 'the biggest civil information technology programme in the world'.[17] As of the time of writing, NPfIT is significantly behind schedule while costs have escalated to about £20 billion. The programme has already diverted money from care for patients, without any assurance that a working setup would be ready on time to service the new market-oriented commissioners/providers split. Again, there was no fallback situation: in an end-state plan the components must come together at the right time and in the right place. What happens if they do not is unthinkable.

Of course the unthinkable does often happen in human activity systems subject to local and external influences by numerous interacting elements aided and abetted by positive feedback, and that happens for perfectly understandable reasons. For example, the Darzi reforms were premised on the continual expansion of NHS budgets that had been going on since 1997. Implementation, however, coincided with the start of the 2008 credit crisis/recession. To make matters worse, the upheaval was unfolding at a time of mounting political activity in preparation for a general election that had to take place by May 2010. At the time of writing it is confidently expected that the next wave of 'reforms' will be launched soon after the election and the process will start again unless a more fundamental change in thinking materialises.

Good and bad news

The good news is that complexity and complexity-like thinking is increasingly penetrating the elite thinking within the English NHS. In addition to growing academic and policy activities and literature,[18] the NHS Institute for Innovation and Improvement launched an Academy for Large Scale Change in 2008 with a clear focus on complex systems as evidenced by the appointment of Paul Plsek, a noted complexity expert, as director. Moreover, soft management systems and lean technology is known and practiced within the NHS in the form of 'production wards', and so on. A number of other ideas have been put forward to provide a strategically higher degree of independence for the NHS.[19] In particular, the transformation of hospitals, and in future community care and primary care trusts, into semi-independent Foundation Trusts (FTs) could be a step in the right direction. If FTs are regulated at arm's length from both the Department of Health and Strategic Health Authorities and develop governance arrangements that give local stakeholders greater powers to offset pressure from government at the national level, then a new more 'complex' system could be in the making.

At the top political level, David Cameron, the leader of the Conservative opposition, announced in a speech to the King's Fund on 9 October 2006 that he would promote a bill to give the NHS greater independence by taking politics out of the NHS. Previously, two minsters in the Labour government also proposed giving a charter to the NHS similar to that enjoyed by the British Broadcasting Corporation (BBC).[20] Most recently, Prime Minister Gordon Brown in his June 2009 *Building Britain's Future* policy document has called for a fundamental rethinking of the dominance of central actors and targeting strategies in the NHS claiming that the state should now empower local actors/consumers to drive forward essential changes. Clearly, this document is a reflection of the new economic constraints on all policy areas and demonstrates a desire by the centre to share out and localise some of the inevitable costs of the current recession. Nevertheless, on the surface at least, UK central actors do seem to be recognising the limits of the orderly paradigm in policy.

Unfortunately, the bad news is that these changes are predominately at the surface level and do not represent a more fundamental shift in policy thinking. What this means is that central political actors in the UK are willing to accept a degree of decentralisation and complexity-like strategies when the situation demands it. However, their underlying position is that as soon as conditions return to 'normal' recentralisation and increased targeting and auditing can occur. The result is a continual pendulum like swinging between these two frameworks. For the English NHS this implies that following a brief period of decentralisation it will soon be buffeted by more 'radical reforms' conceived as fixed plans with numerous elements that must all come together clockwork fashion into a perfectly functioning whole according to a brief timetable determined by political and personal agendas. The associated costs and risks are both obvious and considerable.

The problem is that, as we have seen in so many other policy areas, steeped in the paradigm of order decision-makers at the centre wish/need to demonstrate control and leadership, something that the public at large often encourages them to do. This overweening focus has dire consequences as each incoming administration, minister or senior civil servant seeks to outdo those who were previously in charge. The process is made worse by the 'revolving door' between business and politics. People at the top flit in and out and manage to see only a glimpse of the truly complex system that is the English NHS. Fundamentally, styles of management appropriate for complex systems based as they are on evolutionary change through reiterative process of action, learning and reconsideration contradict the underlying orderly governance model. Treating a system with the high degree of complexity presented by the English NHS as a mechanistic system that could be redesigned by dictate is a hit-and-miss affair. Experience so far indicates it is more miss than hit. A complexity framework may help to make it more hit than miss and calm some of its more extreme 'reorganisations'.

Complexity and chronic illness: the case of diabetes

Now, moving to the other end of the health spectrum, we want to explore how complexity can be used to understand and manage health at an individual level by exploring the case of one of the most widespread chronic illnesses: diabetes.

According to the World Health Organization, the total number of people worldwide with diabetes is projected to rise from 171 million in 2000 to 366 million in 2030 (Wild *et al.* 2004). In the UK, there are an estimate 2 million people with diabetes, while the NHS spends around 9% of its total budget dealing with diabetes and its complications (Currie *et al.* 1997). Reflecting the growing general emphasis on preventative patient education and the continued increase in type 2 diabetes, the NHS has put an increasing amount of money and effort into patient education programmes for diabetes. These programmes have demonstrated that increased contact with health professionals and/or peer group support, alongside changes in knowledge, skills and attitudes generally improves the management of the condition. From a traditional perspective, this implies more education would lead to more improvement and so on.

But will more and longer education (and a lot more resources) really improve diabetes management? Fundamentally, this depends on the framework through which diabetes is examined. If viewed through the lens of the traditional orderly medical framework, diabetes is a complicated condition, but with enough information, effort and patient compliance, the condition can be controlled. From this perspective, poor diabetes management must be the result of not enough education (either in the short or long term) or wayward patients. Hence, the answer is more education and, if possible, more patient compliance.

However, with a complexity perspective there is a structure that incorporates a dynamic, emergent, creative, and intuitive view of the world to replace the traditional 'reduce and resolve' approaches to clinical care and service organisation. Under the lens of complexity, basic education strategies (diet, exercise, etc.) matter, but so do patient interaction with those strategies and the fundamental framework through which patients view their own self-management. In other words, shifting the way that patients and practitioners view diabetes management may be much more important than giving them more of the same.

Understanding chronic illness – from order to complexity

As discussed earlier, the framework of order saw the natural world and human beings as basically orderly and, in many ways, mechanical. In this framework, medicine was about fixing the broken bit of the human 'clock'. The health professional was the expert and the patient the passive recipient of their knowledge. The hope was that as more and more bits of the clock were understood, any broken bits could be fixed and eventually there would be no sickness or disease. Nevertheless, as more and more scientists were realising during the twentieth century, nature and human beings were just not that orderly. In medicine, more and more diseases just wouldn't go away and new ones kept emerging. Traditional health organisations seemed to generate their own negative outcomes and greater central control over health structures was no guarantee of improved health outcomes.

In this context, chronic health issues are ideal for applying complexity concepts and tools. Like other complex systems – environmental conditions, traffic systems, work environments, family relations – chronic illnesses combine predictable and unpredictable elements into an evolving and emergent whole. Diabetes fits easily into this framework. In many ways, getting diabetes is no different from any other long-term alteration in our lives, be they the birth of a child, the pursuit of a new career, or the death of a close relative. They all involve managing and balancing new boundaries with the process of making new discoveries and learning. All of these systems evolve over time and attempting to fix the endpoint of these 'systems' is not only impossible, but frustrating and can be dangerous.

For the person with diabetes, viewing the condition through the traditional orderly framework, it is all about control; taking the right drug doses, structuring lifestyle as rigidly as possible and hoping that a cure may be found in the future. Frustration stems from the continual balancing nature of the disease, where even if you do the right thing (dosages, diet, exercise, etc.) problems and complications may still arise. From an orderly framework, since all problems are basically orderly and patients have been told what to do, any mishaps must be their fault. For the person with diabetes, viewing their condition through an orderly lens, guilt can be a frequent companion.

For the health professional, steeped in the belief that it is just a matter of effort and control, they are condemned to play the role of a nagging teacher. Patients should listen and comply if they are to get better. Patients who do not manage their condition properly are seen as wayward, not paying attention and not following the rules. For these cases, more monitoring and pressure are necessary to bring out better behaviour. The longer the patient shows a reluctance to self-manage effectively, the greater the pressure that must be applied.

Obviously, this is not true in all cases! But the pressures remain nonetheless. Over time, it has become increasingly obvious that an orderly framework for diabetes just does not fit. In the last 10–15 years the tide has begun to turn. Patient education strategies have been developed based on evidence-based guidelines that reflect the importance of self-managing diabetes and of listening to, rather than talking at, patients.[21] This healthy change is not a direct product of complexity thinking, but reflects the general shift away from the orderly framework that is going on throughout the fields of science, medicine and public policy.

Whilst these recent changes have been welcomed, the question remains whether they are enough? Carers and patients are being taught diabetes self-management strategies, but fitting them all together on a day-to-day basis within the real world often remains a challenge. Without a framework to guide this, there will be a natural tendency to drift back towards old ways, and the need for more education reflective of teacher–child relationships. This regressive tendency is strengthened by the growing command-and-control 'audit culture'. As increasing pressure is put on health professionals to show results (meet targets!) for their interventions, they will obviously be encouraged to put more pressure on patients to get these results. And before you know it, we are back to where we started!

The best way to stop this wasteful and repetitive mistake is to shift the scientific framework that underpins these strategies. With the complexity framework patients and practitioners can begin to develop a very different philosophical approach to self-management; one that addresses the many different interconnecting parts of a long-term condition like diabetes and the emergent nature of learning how to live with the condition.

Using a fitness landscape to improve diabetes management

In this brief section we can only explore one small tool of complexity, the difference between a more linear x–y graph and the more complex fitness landscape.[22] As demonstrated in Chapter 3, a classic mental model from an orderly perspective is the standard x–y graph. With an x–y graph, any point on a two-dimensional surface can be plotted and connected. For orderly-based thinkers, the person with diabetes and process of diabetes could be represented using Figure 5.1.

From an orderly mechanistic perspective, the approach to the person with diabetes is to get them to move from point A to point B as quickly as possible and then to keep them moving along from point B to point C for the foreseeable future. This type of perspective is exactly what many policy actors want to see because it is easy to follow the progress of any given patient, calculate where the average patient is on the graph, and judge the efficiency of any given

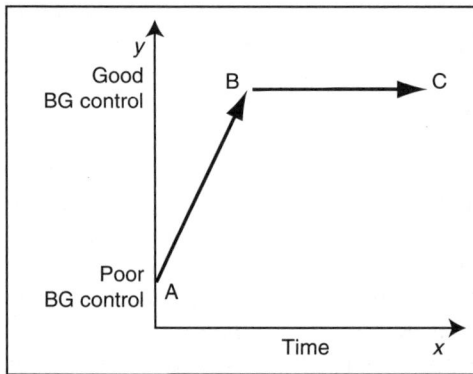

Figure 5.1 An orderly perspective: the x–y graph and blood glucose control.

hospital or treatment centre. However, this perspective has several disturbing implications.

- The faster one gets a patient to B and the more the patient stays on the line between B and C, the better the strategy must be. Hence, this strategy should be best for other patients. All patients are alike.
- Once the patient is on the B–C line they are encouraged to do exactly as they have done before as this is the most likely strategy for keeping them on the line. Diabetes management is basically repetition. If something goes wrong, repeat previous successful strategies. Experimentation, exploration and learning are best avoided and adopted only as a last resort.
- Patients, due to their lack of knowledge, must be fundamentally passive, rely on experts and do as they are told.
- Failing to move quickly from A to B or falling off the B to C line equals a 'lack of control' or failure to follow the rules. It is intrinsically the patient's fault.

The image that the x–y graph conjures in the mind is that diabetes is like walking up a steep mountain. The expert is there to guide the patient up the mountain of control as quickly and efficiently as possible. Once the patient is at the top, the peak of control, there is nowhere else to go except to hold still and repeat what the expert said. Any movement off the peak implies failure, mistakes, and/or lack of control. The anxiety that is created by this desperate struggle to stay still through the normal bumpy ride of life is obviously enormous.

A different point of view: the fitness landscape

For simple orderly phenomena and systems, the x–y graph is an excellent tool. To calculate the movements of balls on a table, and so on, the x–y graph is all you need. However, for complex systems you need a way of modelling and imagining a system that can move in varying and unpredictable ways over time. You need a way to show the probability of a system to move in a multitude of directions, a way to show how the system 'fits' with varying situations and circumstances. You need a 'fitness landscape'.

As we saw in Chapter 3, a fitness landscape for a given system, the patient in this instance, is a three-dimensional representation. Different locations on the x–y plane demonstrate the wide variety of conditions that the system (individual) might have to negotiate. The height/depth on the x axis represents levels of fitness. Valleys in the landscape represent areas of poor fitness, mountains represent high fitness and flatlands represent areas of neutral fitness. The challenge for the system/individual/patient is to maximise its fitness under a wide variety of conditions. Some of the basic rules for survival on a fitness landscape are adaptability, flexibility, learning and balance. As long as the 'system' survives, the fitness landscape continues to evolve through time like a never-ending conveyor belt.

With regards to diabetes, imagine a landscape that is full of flatlands, valleys and mountains that stretches endlessly into the future. Now, imagine that the valleys represent zones of poor control, the mountains are zones of good control and the flatlands are areas of uneven control. Notice that the mountains and valleys are not evenly distributed. A number of mountains, flatlands and occasional valleys are clustered straight ahead of you. On either side, the mountains begin to shrink and become less common, the flatlands are fewer, and the valleys are more numerous and deeper.

This is a good image of the fitness landscape for the average person with diabetes. Straight ahead is the path of creative complexity: a mixture of reasonable basic strategies (good diet, regular exercise, balanced insulin usage, etc.) and continual learning (exploring what types of food, exercise and strategies suit the individual). This zone has many mountains of good fitness, but these don't stay the same all of the time. They are clustered in the reasonable strategies zone because these strategies are the most likely to produce good balance, but hidden valleys abound. A stressful period, illness, hormonal changes, pregnancy, and other significant life events all mask hidden valleys that a strict adherence to previous strategies might not solve. On either side of the zone of reasonable strategies lie increasingly uncertain areas of deep and frequent valleys of poor balance and only occasional mountains. As this image implies, life is tougher in these areas. For most people with diabetes, good balance is tougher to maintain in these zones, but mountains of good balance do exist. Alcoholics or smokers may choose to live out in these more difficult zones. People with unstable or so-called 'brittle' diabetes are often forced to travel through them. Even those who try to rigidly follow the advice of their doctor (structuring their lives completely around the condition) may unknowingly be travelling along a more difficult orderly terrain.

Just as species evolving through time, the primary tactics of people with diabetes travelling through their fitness landscape are adaptability, flexibility, learning

108 *Health*

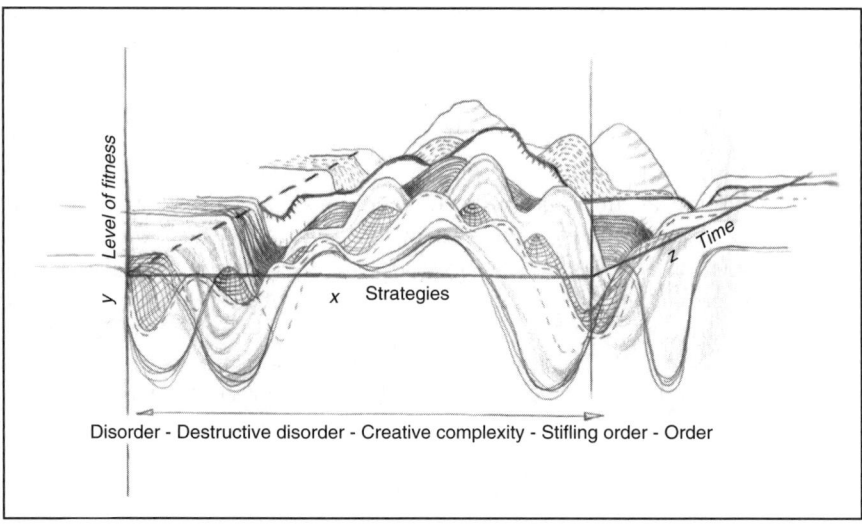

Figure 5.2 Visual representation of a fitness landscape of strategies for managing diabetes. Drawing created by Laura Fleming, http//laurafleming.co.uk/html/home.htm.

and balance. As opposed to our orderly vision of a walk to the mountaintop of control, a trek through the fitness landscape of diabetes reveals a number of remarkable 'common sense' implications:

- There is no endpoint to a fitness landscape nor is there any final resting point. The primary goal and tactic is adaptation and balance to changing circumstances.
- The main actor on treks on the fitness landscape is the patient, actively moving through the landscape and making essential choices, where personal opinions, experiences and learning matter.
- Learning is essential and never stops. Being aware, making choices, experimenting, exploring is how the patient learns about their landscape and evolves personal tactics for dealing with the inevitable unknown. Health professionals are helpers on the journey. Their advice is important, but ultimately the patient decides.
- Change is not bad. Patients are not desperately struggling to cling to a mountaintop hoping that the next wind won't blow them off. They are learning a new way of living that offers a whole range of problems and challenges. Change is part of that world. Exploring and accepting that change is what makes it all worthwhile.
- As the patient explores, mistakes, mis-directions and occasional stumbles into valleys are *normal* and not a mark of weakness or failure. The only mistake is refusing to see them as normal and not learning from them.

Bringing it all together

In this chapter, using global, institutional and individual level views of health we have tried to demonstrate how the orderly paradigm has structured our understanding of health at all levels of society. The way patients view chronic conditions, health professionals treat patients, hospitals are organised, national health systems are structured and global health policies are developed all reflect the legacy of order. We did not have the space to explore all of these aspects within this chapter. Nevertheless, we believe that the implications of complexity for health are clear:

- Local actors (patient, localities, hospital, etc.) matter. Health systems and individual health run on a vast number of daily micro-decisions made by local actors – patients, carers, nurses, and so on.
- Focus on the basics and concentrate on developing local coping strategies. For patients this means emphasising individual responsibility and good health practices over strict 'patient compliance'. For poorer nations it implies integrating good basic strategies with local knowledge and conditions.
- More money and/or more oversight will not automatically improve health outcomes for individuals, institutions or policy areas. In particular, if the system (patient or policy) remains constrained within an orderly framework, more doctor supervision or managerial oversight of the policy may actually lead to increasingly negative outcomes.
- Health is always about 'balance'. Individuals and societies could always have more health. There is no limit to the money, time and effort that could be spent to promote it. Moreover, there is no endpoint to health or for that matter perfect health. Balancing health against other individual and societal needs is a continuing process.
- Different 'tools' can help you to see your own health and your nation's differently. In the chapter we looked at the implications of using a fitness landscape to understand diabetes management. However, other complexity 'tools' and concepts (complexity cascade, complexity mapping, Stacey diagram, etc.) are equally relevant to the individual as well as local, national and international policy-makers.

6 The international arena

> May I just clarify the question? You're asking who would know what it is that I don't know and you don't know but the Foreign Office know that they know, that they are keeping from you so that you don't know but they do know, and all we know is that there is something we don't know and we want to know but we don't know what because we don't know. Is that it?[1]
>
> (From the 1980s British TV Programme, *Yes Prime Minister* – Bernard, the assistant to Prime Minister Hacker, is trying to clarify to the PM what the British Foreign Office knows about a potentially embarrassing political crisis).

> You are either with us or against us.
>
> (President George Bush Jr.)

Which of these quotations most accurately captures the nature of the international arena? On the one hand, the chaotic and confusing nature of the international system seems blatantly obvious. Numerous actors playing power games on multiple levels evolving around national, regional and global dynamics indicate such a high degree of complexity that one's head begins to spin. On the other hand, the remarkably stable international power inequalities, centrality of economic and military power and willingness of key actors to see the world in simple divisions (good–bad, anti-communist–communist, anti-terrorist–terrorist) indicate that there are fundamentally orderly elements to the international system. Not surprisingly, theories of the international realm reflect this disorderly/orderly division.

In this chapter we will explore how complexity provides a new way of comprehending the international realm by examining how it can be used to understand the evolution of international relations and European integration and by looking at a couple of interesting case studies: the rise and fall of the concepts of 'globalisation' and 'Europeanisation', the curious non-death of Scandinavian exceptionalism and the call for a true capital of Europe.

International relations theory

As is well known, international relations (IR) in the early post-WWII period was dominated by the theory of realism (Knutsen 1997). Realism assumed nation-states were the primary units at the international level They were rational utility-maximisers and the international level was an amoral anarchical arena where nation-states competed against one another for economic, political and military advantage. In essence, the system had a clear unchanging order (states in anarchy) that unsurprisingly reflected the experience of the Cold War. Given these assumptions, the international system could be understood from a positivist epistemological and methodological perspective. Nation-states behaved like balls in motion on a pool table and their behaviour and capabilities were assumed to conform to Newton's laws of motion. They could be rationally calculated and predicted and would tend towards equilibrium (the 'balance of power' concept). Prediction and control were obviously essential in a world where nervous fingers were poised on nuclear triggers.

By the 1970s with the collapse of the Bretton Woods economic system, growth of transnational corporations and cooling of the Cold War, interdependence or regime theorists began to emerge (Keohane and Nye 1997; Krasner 1983). They stressed that the international system was not wholly anarchical, international actors had emerged and were increasingly important and the actions and interests of national actors could be reshaped by the 'web of interdependence' or 'regimes'. These theorists often tried to adhere to the orderly tradition. However, the 'bounded rationality' of the main actors, the growing number and complexity of key actors and uncertain developments made a strict adherence to this tradition increasingly difficult. The international arena could no longer be understood as uniformly orderly and therefore analysed through purely reductionist and parsimonious strategies.

In the 1980s and 1990s both realists and interdependence/regime theorists were criticised by 'reflectivist' theories (Ashley 1986; Checkel 1998; Walker 1993). Emerging from a broad range of positions from critical theory and feminism, to post-modernism and post-structuralism, reflectivists emphasised that much of IR (and realism in particular) were ideological constructs created by the dominant powers in the international system. Neither the actors nor the system were inherently rational and what was deemed to be rational in one time or context may vary in another time or context. Many reflectivists adopted anti-naturalist and anti-foundationalist positions, arguing that human experience was inherently distinctive from natural phenomena and that there could be no certain epistemological foundations for claims to fundamental human truths. Reality (or anarchy as one author argued) was what one made of it.

From the early 1990s, 'constructivists' (Adler 1997; Christiansen *et al.* 2001; Onuf 1989; Wendt 1999;) attempted to 'build a bridge between these two traditions' (Wendt 1992: 394) by emphasising ontological and epistemological openness. Not surprisingly, despite these bridge-building efforts both rationalists and reflectivists have continued to exclude and ignore each other while clinging to the certainty of their orderly or disorderly ontological/epistemological claims. For example, rationalists attempted to co-opt constructivism by arguing that 'rationalism ... and constructivism now provide the major points of contestation for international relations scholarship' (Katzenstein *et al.* 1998: 646) and exclude reflectivism by stressing that:

> [it] denies ... the use of evidence to adjudicate between truth claims ... [it] falls clearly outside of the social science enterprise, and in IR research it risks becoming self-referential and disengaged from the world, protests to the contrary notwithstanding.
>
> (Katzenstein *et al.* 1998: 678)

On the other side, reflectivists have complained that social constructivism goes too far in a rationalist direction, accepting many of the major constructs such as the primacy of nation-states and drifting towards a positivist methodology (Smith 2001). Therefore, is the bridge-building strategy of constructivism doomed to failure? How can international relations thinking go forward from here?

European integration theory

European integration (EI) theory mirrored much of the post-WWII development of IR theory (Chryssocyoou 2001; Rosamond 2000). In the 1950s and 60s, the core European integration debate involved intergovernmentalists, who saw the EU as an intergovernmental extension of a fundamentally realist international order, and functionalists/neo-functionalists, who saw the early EU as possessing the ability to functionally reshape the realist international order (at least within Western Europe). During these years debates raged over the degree to which early EU policy developments were determined by intergovernmental bargains or functional spillover. The fates of the theories were tied to the success or failure of the integration process. When it succeeded, neo-functionalists boasted. When it faltered, intergovernmentalists exulted.

Following a period of stagnation in the 1970s, when many integration theorists drifted to other areas of research,[2] integration theory revived in the 1980s and 1990s with the revival of integration through the Single European Market project. New theories, linked to the earlier ones, began to recognise the more complex and uncertain nature of European integration (Taylor 1983). Andrew Moravcsik carried the torch for intergovermentalists. However, even his concept of liberal intergovernmentalism recognised the importance of complex institutional dynamics (Moravcsik 1993). Others held on to a modified neo-functionalism (Tranholm-Mikkelsen 1991). Both theories were brought together by multi-level governance theorists (Bache and Flinders 2005; Hooghe and Marks 2001) who argued that the EU was composed of 'overlapping competencies of among multiple levels of governments and the interaction of political actors across those levels' (Marks *et al.* 1996: 41).

Despite this increasing recognition of complexity, or because of it, reflectivist and constructivist works came late to EI theory, only beginning to emerge in the late 1990s (Christiansen *et al.* 2001; Checkel 1998, 1999; Diez 1999; Jørgensen 1997). Again, similar to the experience in IR theory, constructivists saw themselves as 'establishing a middle ground' (Christiansen *et al.* 2001: 8) between rationalist and reflectivist paradigms. Unsurprisingly, they came under fire from both sides of the debate. On the one hand, reflectivists complained that it was:

> far more 'rationalist' in character than 'reflectivist'; indeed I would go so far as to say that social constructivism in its dominant (mainly North American) form is very close to the neo-liberalist wing of the rationalist paradigm.
> (Smith 2001: 191)

On the other hand, rationalists argued that:

> All this (philosophical speculation) distracts constructivists from the only element truly essential to social science: the vulnerability of conjectures to some sort of empirical disconfirmation.
> (Moravcsik 2001: 186)

Moreover, Mark Pollack, echoing the conclusions of Katzenstein *et al.* (1998) in IR theory, argued that EI theory must accept 'broader ontologies', but:

> We must necessarily fall back on careful empirical testing ... as the ultimate, and indeed the only, standard of what constitutes 'good work' and what constitutes support for one approach or another.
> (Pollack 2001: 236)

Just like IR, EI theory was divided into two opposing poles and a struggling bridging strategy.

Integrating complexity

What does this brief review of IR and EI theory demonstrate? First, there has been a significant challenge to the hegemonic position of the rationalist paradigm in IR and EI theory since the 1970s. Second, linked to this challenge has been the growing recognition of human and social complexity. Third, a core division has emerged within the discipline between rationalists who adopt a strong naturalist position, modelling themselves on a traditional view of the natural sciences, and reflectivists who adopt an anti-naturalist position and oppose the use of natural science epistemologies and methods in the human sphere. Lastly, constructivists have attempted to bridge this division by emphasising the importance of broader ontologies, but have been rejected and/or co-opted by both sides.

How does complexity theory fit into these debates? Unsurprisingly, the growth of complexity in the social sciences has begun to spill over into IR and EU theory.[3]

114 The international arena

Kavalski (2007) goes so far as to claim that the growing influence of complexity thinking in IR and EU theory represents a fifth major debate in the history of these fields. As we have seen, complexity theory argues that order, complexity and disorder all play a role in the creation of the natural and human world. For complexity theory, there are orderly, complex and disorderly phenomena and different epistemological and methodological strategies apply to each. Universal laws and order only apply to certain phenomena. This implies that the fundamental naturalist–anti-naturalist division within the IR and EI theory is based on an out-of-date view of the natural sciences. The natural sciences have not stood still. They have gone through a Kuhnian paradigmatic shift that challenges the traditional naturalist–anti-naturalist division. Without this division, neither rationalists nor reflectivists can claim to have a superior grasp of reality or a greater access to the 'truth' since both are only describing part of the picture and 'the divisions between "rationalist" and "reflectivist" ... will become progressively harder to draw' (Rengger 2000: 195).

Applying a complexity map to the international arena and European Union

The easiest way to view the international arena as a complex system is to insert it into the complexity mapping tool developed in Chapter 3. This is demonstrated in Figure 6.1.

The short term basic framework of the system, particularly the existence of stable nation-states and their significant power inequalities, appears as its most obvious orderly aspects. Nation-states have been a stable element of the international system for hundreds of years. Meanwhile, the power of the US significantly influences its range of options for responding to events such as those of 11 September whereas the power of Belgium, for example, to respond to those developments is much more limited. Dynamics similar to physical complexity can

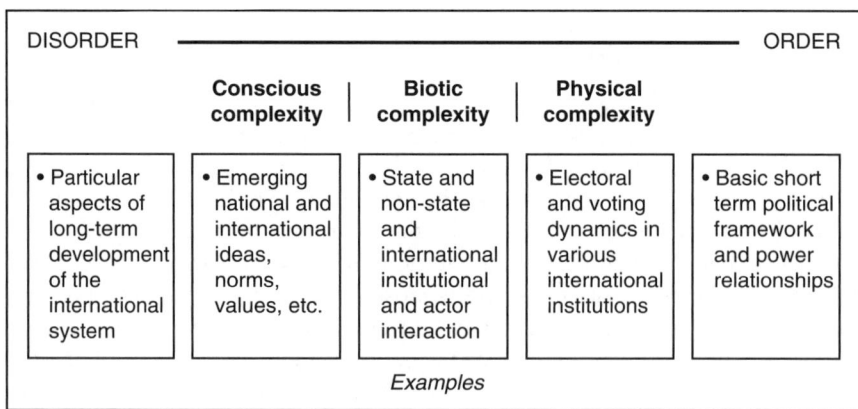

Figure 6.1 The range of international arena phenomena.

be immediately perceived within the procedures of the international institutions and regimes (UN, regional and trade organisations, etc.) which pervade the international system. The voting pattern in the UN of supporting the US in its 'war on terrorism' led to a recognisable pattern of responses from various countries. However, the exact reasons behind these decisions or whether an exact group would support a later 'war on terrorism' would be extremely hard to predict.

Dynamics similar to those found in biotic complex systems can easily be identified in the adaptive and interactive strategies that emerge when international institutions, states and non-state actors interact with each other. For example, as the US began talking about expanding the 'war on terror' to include other nations (North Korea, Libya, etc.) they upset the balance in the coalition which was supporting the actions in Afghanistan. Allies began to weaken their support and question other areas, Palestinian conflict in Israel, of US international policy. This was obviously complicated by the conscious complexity of competing norms and interpretations of the 'war' and US policies. A good example would be the different interpretations of the treatment and rights of the prisoners at Camp X-Ray at Guantanamo Bay. Finally, the exact long-term development of the international system would seem to be the most uncertain and unpredictable analysis. How could an observer of the international system of 1900 have predicted the rise of communism, two world wars and the hegemony of the US by 1950? How could an observer in 1950 have predicted the economic success of the West, collapse of communism and rise of the European Union by 2000? The range of possible developments and interactions is enormously large. In a similar fashion, Figure 6.2 shows that the European Union can easily be inserted into a complexity map.

The most orderly aspect of the EU is its core short-term framework. There is a very high degree of probability that the basic structures of the EU (core member-state voting balances, institutional structures, power relationships) will remain stable in the short term. The various voting procedures in the EU Parliament and Council demonstrate aspects of physical complexity of the EU. Choices are

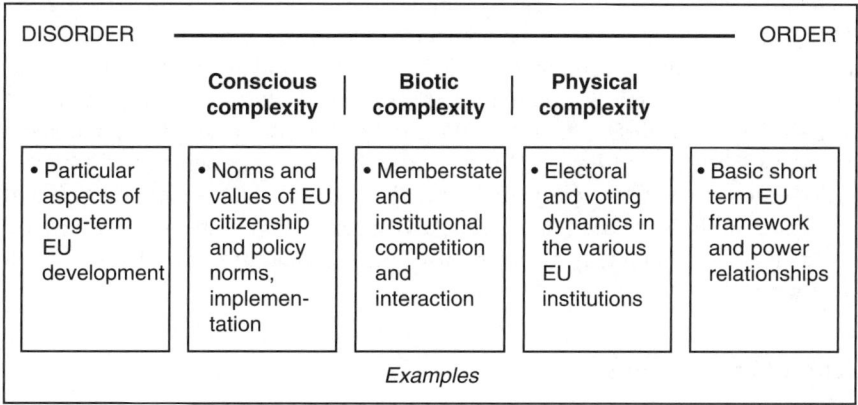

Figure 6.2   A complexity map of the European Union.

constrained by having to vote for or against a particular proposal and may produce stable patterns (certain member-states or groups always voting for/against certain proposals). However, the pattern is not continuously stable, nor can one be certain that similar voting patterns at different times were based on the exact same factors. For example, the Scandinavian countries may generally vote as a bloc on most proposals, demonstrating a stable pattern, but their exact reasons for doing so may vary substantial with every proposal.

As any multi-level governance theorist would point out, aspects of biotic complexity are most obvious in the multiple types of member-state and EU institutional interaction. Member-states and EU institutions are constantly interacting in evolving and adaptive ways to new policies and developments within the EU system. Stable patterns emerge, but these are much more susceptible to unpredictable developments as different member-states and institutions constantly evolve and adjust to new opportunities and constraints. For example, many observers of EU social policy expected it to expand rapidly after the defeat of the British Conservative government in 1997. However, despite a more receptive pro-social policy Labour government other supposedly pro-EU social policy member-states suddenly became less supportive when they realised that their social policy proposals would no longer be automatically defeated by a British veto. The basic pattern of limited EU social policy developments continued, but the internal dynamics were transformed (Geyer 2000).

Conscious complexity is most obvious in areas such as the meaning of EU citizenship and the implementation of EU policies in the member-states. EU citizenship is an extremely contested concept and means drastically different things to the various member-states and groups within the member-states. This is complicated by the continual evolution of the norms surrounding 'citizenship' within the member-states themselves and how the actions of the EU, its policies, are implemented and interpreted by the member-states and their populations.

A good example of the disorderly nature of the EU would be assumptions of its long-term development. There are so many possible outcomes of the mechanical, organic and conscious complex systems with and surrounding the EU that predicting its exact development in the long term is obviously nearly impossible. Imagine a lecturer trying to tell his/her students at the founding of the European Economic Community (EEC) in 1957, during the early phase of the Cold War, that in 50 years time the EEC will grow from 6 to 27 members (including Western and Eastern Europe) and that half the countries will have a unified currency called the euro. Such a prediction would have seemed absurd and the lecturer might have been fired. At best, one can guess or pick a future that one would like to see, but it may have virtually no direct relationship to the one which will emerge.

General implications

If the EU and international system can be interpreted as complex systems, then there are several major implications. First, this implies 'incompressibility', that any truly accurate description must be as detailed as the system itself. Hence, the pursuit of parsimonious order must of necessity take the observer further and further from holistic reality. Second, since the EU and international systems are composed of different phenomena then one must accept methodological pluralism; quantitative modelling, qualitative analysis, historical description and narrative discourse all have their place with regard to particular phenomena. There is no universal hierarchy of phenomena and hence no hierarchy of method. Third, this uncertainty or perceived lack of knowledge is actually a strength of human systems since:

> it is this very 'ignorance' or multiple misunderstandings that generates microdiversity, and leads therefore to exploration and (imperfect) learning.
> (Allen 2001: 41)

This means that some degree of different interpretations, diverse interests, uncertain responses, clumsy adaptations, learning and mistakes are what keep a system healthy and evolving. Truly orderly systems, where all of the elements are at the average, are dead systems and have no ability to explore new patterns or adapt to new environments.

The EU itself provides an excellent example of the healthy nature of complexity. From an orderly rationalist framework it is incomprehensively messy. From a reflectivist standpoint it cannot possibly order the multi-faceted and multi-level nature of its constituent societies and sub-groups. Nevertheless, it exists and thrives by providing a reasonable and pragmatic framework of institutions for promoting complex interaction, learning, diversity and adaptation at the sub-national, national and European levels. Just imagine how long the EU would last if it did try to assert a comprehensive rigid order on the multitude of member-states. Even its most rigid policies, such as Economic and Monetary Union, allow for a surprising amount of hidden flexibility and adaptation. In fact, it is the very flexibility of the other aspects of economic policy that allows the member-states to accept the rigidity within the European monetary order. As Hodson and Maher explain in relation to the EU's open method of coordination (developed for the 2000 Lisbon European Council) for promoting economic policy cooperation:

> This is no formal attempt to control outcomes (outside of fiscal policy of course), and process is determined by a system of benchmarking and lesson-drawing, emphasizing state competence and the voluntary alignment of policies ... The desire of the EC to control outcomes, as manifest in the directive as the rule of choice in the single market, with its emphasis on common outcomes if not methods, is overcome by recognition of the importance of diversity at the national level in relation to policy formation, legal frameworks, ideational references and popular perceptions and reactions to either the European project generally or the specific policy being co-ordinated.
> (Hodson and Maher 2001: 731)

118 *The international arena*

In essence, the EU works because it combines an agreed fundamental framework with member-state diversity and autonomy. This was not based on any preordained plan, but emerged from a multi-faceted combination of historical events and political economic structures including the weakness of the EU as a power centre, the continuing resilience of the member-states to oppose centralising EU initiatives and the evolving nature of the international system. A significant change in any of these factors could easily have disrupted the reasonably complex development of the EU.

Globalisation, Europeanisation and the curious case of the non-death of Scandinavian exceptionalism

Given that complexity has implications for IR and EU theory, what can it tell us about how we view other aspects of the international realm?

Globalisation

Emerging out of the rapid development of international capital markets in the 1970s, 80s and 90s, the revival of neo-liberal economic policy in the US and UK (Reagan and Thatcher revolutions), and the economic difficulties following the collapse of the Bretton Woods system, globalisation was seen as a new hegemonic economic force which would empower capital, undermine state powers, and force all advanced industrial countries (let alone Third World countries) to pursue neo-liberal economic policies, abandon welfare states and create a destructive competition between national social and environmental systems of regulation. A huge academic and non-academic literature emerged at this time, the zenith of this type of thinking was epitomised by the work of Kenichi Ohmae, *The Borderless World* (1994) and Francis Fukuyama (1992) who saw globalisation fundamentally undermining nation-states and locking into place a new neo-liberal world order.

However, as political and academic debates raged in the 1990s, it became increasingly obvious that the impact and development of globalisation was more complex than the early thinkers/ideologues had assumed. Summarised nicely in the work of Paul Hirst and Grahame Thompson (Hirst and Thompson 1996), observers began to note that despite growing economic regionalisation within the wealthy countries, poorer societies were being left out of the process. Despite greater capital mobility and the internationalisation of production, general trade flows and patterns remained remarkably stable. Despite the collapse of traditional Keynesian fiscal policy, active monetary, regional, and labour market policies remained viable. Finally, despite significant pressures on taxation levels and welfare regimes in advanced industrial states, taxation levels had remained remarkably stable (Swank 1998) and welfare state expenditure had actually grown slightly during the period (Hay 2001). Not surprisingly, given this growing body of 'limited globalisation', the focus began to shift from seeing globalisation as a linear and homogenising 'hegemonic force' to a more 'interactive influence' on national systems. By the end of the 1990s, globalisation became much more uncertain, variable, complex and interdependent. In the 2000s, the 'myth' of globalisation was being attacked from a number of directions (Held and McGrew 2007; Scholte 2005; Stiglitz 2003).

Europeanisation: the baby brother of globalisation

Europeanisation as a marketable academic concept emerged on the back of the aforementioned success of European integration and impact of globalisation in the 1990s. For many, particularly in the early 1990s, it became the regional extension of the globalisation debate. Since the international economy was globalising, the EU had to either embrace and enhance this development or build walls to protect the distinctive importance of 'Social Europe'. These debates were particularly visible in the areas of EU social policy and European welfare-state research (Geyer 2000; Leibfried and Pierson 1995; Rhodes 1996). However, again similar to the fate of globalisation, Europeanisation became a much more subtle, complex and interactive concept as the 1990s and 2000s progressed. Despite the growing influence of the EU in a multitude of policy areas, national policy regimes remained remarkably distinctive. The research focus began to shift from how Europe was shaping national policy regimes to how national regimes were interacting with and adjusting to EU developments. Numerous nationally and comparatively oriented works began to emerge to explore this detailed interaction (Bache and Jordon 2006; Bonoli *et al*. 2000; Ferrera and Rhodes 2000; Esping-Andersen 1996; Geyer and MacIntosh 2005; and Sykes *et al*. 2001). In essence, in the new millennium, simple positions regarding the costs and benefits of the EU and Europeanisation were being buried under a mountain of more subtle, interactive and complex analyses.

The curious case of the non-death of Scandinavian exceptionalism

What did this earlier type of orderly thinking about globalisation and Europeanisation lead many to conclude regarding the fate of individual nation-states? An intriguing example is the case of the tiny Scandinavian countries and what happened when they were confronted with the forces of globalisation and Europeanisation.

Throughout much of the twentieth century most observers have seen the three Scandinavian countries of Denmark, Norway and Sweden as pursuing a distinctive third or middle way between Anglo-American free market capitalism and Soviet communism. As early as 1936, an American journalist published an amazingly popular book praising the Swedish 'middle way', which went through three editions, fourteen printings and generated an American congressional study on the Swedish co-operative movement (Childs 1936). In more recent times, following the economic difficulties of the 1970s and the political resurgence of the Right in the 1980s, observers (particularly on the Left) increasingly turned to Scandinavia for proof of the viability of social democracy in an increasingly difficult international economic context. Key academic works by leading Scandinavian experts such as Gosta Esping-Andersen's *Politics Against Markets* (1985), Peter Gourevitch's *Politics in Hard Times* (1986), Peter Katzenstein's *Small States in World Markets* (1985) and Walter Korpi's *The Democratic Class Struggle* (1983) all emerged in the mid-1980s. To varying degrees all argued that the Scandinavian countries had been capable of pursuing and maintaining their distinctive social and economic trajectories throughout the turbulent 1970s. The key to this success, depending on which one author selected, was a combination of key social alliances, democratic corporatism, flexible institutional structures, flexible incremental adjustment to international developments, and/or the continued political success (and stability) of the Left and trade union movement.

In the early 1990s, globalisation and Europeanisation replaced the earlier fears of international economic turbulence as the main threat to the distinctive social democratic structure of Scandinavian economic and welfare state policy (Geyer 1997). However, once again, by the mid-to late 1990s, it was becoming increasingly obvious that the distinctive social democratic Scandinavian polities and welfare states continued to survive and adapt to their new international conditions. By the late 1990s and early 2000s several works emerged which detailed the successful adaptation of the Scandinavian countries to the complex and interactive nature of globalisation and Europeanisation (Einhorn and Logue 2003; Geyer, Ingebritsen and Moses 2000; Hanf and Soetendorp 1998; Kuhnle 2000; Miles 2005). Time and time again during the twentieth century, the small and relatively weak Scandinavian social democracies managed to quietly adapt and adjust their systems to fit the new surroundings.

Implications

What conclusions can be drawn from the obstinate refusal of the Scandinavian countries to conform to their orderly fate? First, deterministic international concepts make good book titles (just do a search of "globali(z)sation" on amazon.com) but poor long-term concepts. As discussed above, the earlier interpretations of globalisation, Europeanisation and Scandinavian exceptionalism were based on a orderly framework. If globalisation and Europeanisation are hegemonic structures, then all countries should converge and Scandinavia must lose its exceptionalism. However, as recent work demonstrated, globalisation is not hegemonic and Scandinavia is not losing its distinctiveness. Moreover, with historical hindsight one can easily see that throughout the twentieth century Scandinavia has always been a problem to those who were determined to impose a particular order on the world. The problem is not globalisation, Europeanisation or Scandinavian exceptionalism; it is the belief in and pursuit of a single final order. Peaceful, consensual and economically successful Scandinavian social democracies that mixed elements of both state and market 'orders' were always a conundrum for the traditional model.

What made the work of Katzenstein, Esping-Andersen, Gourevitch, Korpi and others in the mid-1980s so useful and influential was that in trying to confront this conundrum they did not assume that Scandinavia would necessarily have to conform to a given order. Similarly, public policy debates were not to be constricted to concepts of 'plan' *versus* 'market' which represented 'the Peter Pan approach to public policy: one closes one's eyes and wishes really hard' (Katzenstein 1985: 210). By allowing for this openness in their own thinking they indirectly challenged the intellectual limitations set by the Newtonian vision of global human order. With this openness and a strong comparative approach,[4] they rejected simplistic models and performed the difficult academic 'footwork' of tracing the unique historical developments, distinctive institutional mixes, co-operative cultures, evolving social and political alliances and a number of other complex and contingent factors that made Scandinavia unique. This depth, complexity and subtlety led them to refuse neat, orderly and universalistic conclusions. There was no simple lever to pull or policy to pick. In the end, small states were successful, 'because they have found a way to live with change' (Katzenstein 1985: 211).

From a traditional orderly perspective this type of analysis always appeared to be weak and 'soft'. However, as we have seen earlier, from a complexity perspective it makes perfect sense. Complexity does not assume that there is one type or model of development, but a multitude of types and models. Hence, there is not one type of capitalism, but a multitude of types (American, Japanese, German, French, etc.). There is also not a single strategy or 'way' of dealing with economic crises and difficulties, but a multitude of strategies that may be specific to particular contexts and vary over time. Moreover, due to the contingent complex and interactive aspects of socio-economic development, it is nearly impossible to directly copy specific forms of economic adjustment and development. Countries, institutions and individuals can learn from others. However, the learning process is not orderly, but complex and interactive. Policies are never directly transferred, but are transmuted through multiple national and institutional lenses.

On all of these aspects, the 'flexible adaptation' writers of the mid-1980s were pushing the envelope of the orderly paradigm. Complexity theory would agree with them, but push to go a step further. For example, Scandinavian countries are only a conundrum or exceptional from an orderly perspective. If one expects variation and complexity, they are seen as healthy societies exhibiting all of the normal aspects of local interaction, learning, adaptation and balance in an evolving social framework. Further, a complexity perspective would agree with the importance of flexible and adaptive public policy, but argue that there is no clear policy answer to all situations. Beyond creating a stable fundamental order within which individuals can learn, interact and adapt, all policies become probablistic. Policy elites and academics can only know so much. They must constantly learn, adapt and evolve at the same time as balancing uncertainty and intuition against evidence and current practice.

The Scandinavian social democratic societies have much to be proud of. The combination of economic success, political stability, high levels of equality and social inclusion, extensive and universalistic welfare states and other factors make them exceptional countries. This persistent 'exceptionalism' has enabled them

122 *The international arena*

to play a significant international role as an example against visions of a rigid free market order. From the 1930s and 1940s when Childs was popularising the Swedish 'middle way' to the mid-1980s when Katzenstein and others were stressing their abilities of socially co-operative 'flexible adaptation', Scandinavia has challenged the orderly foundations of capitalist and communist theoretical visions. With the collapse of Soviet communism, the spectre of a unifying global paradigm reasserted itself through concepts such as globalisation and Europeanisation and the return of the belief that we had (again!) reached 'the end of history'. However, with the collapse of 'hard' globalisation and Europeanisation and the growth of a complexity framework, Scandinavia might finally lose its status as an exception. As the Scandinavian countries demonstrate, diversity, not order, is normal.

Another example: the predictable proposal for a true capital for Europe

Why is the pursuit of order a continual threat to the EU? Surely a messy institution like the EU could do with more order and less confusion! True, but the EU is constantly confronted with strategies and proposals modelled on a traditional orderly interpretation of nation-state structures that cannot possibly fit its unique development. An excellent example of this type of thinking emerged during the summer of 2001 when the EU Commission President Romano Prodi initiated a new discussion on how to make Brussels a true capital for the EU.[5] With European monetary union in full swing and new members queuing up to join, Prodi clearly felt it was time to make the EU a 'normal' state. The underlying position was that if the EU was going to be a true nation/power it had to follow the pattern of other nation/state and have a true capital. From an orderly point of view this was very reasonable and utterly predictable. A certain stage of development had been reached and obviously appropriate institutions had to be created.

From a complexity perspective, trying to force Brussels to be seen as and act like a traditional national capital for Europe would be absurd and dangerous to the European project.[6] A quick glance at the history of the development of European national capitals demonstrates that their evolution was not particularly glorious and more often repressive. They were mostly built by non-democratic monarchies and dictatorships whose primary purposes were the centralisation of national power, repression of regional and cultural variations and the imposition of domineering regimes of national unification. Their importance was demonstrated by their military value while their art and architecture often exhibited the most blatant anti-humanist styles from neo-classicalism to socialist realism.

The European Union was not, nor is likely to become, a 'traditional' nation-state. The EU was not designed nor intended to completely replace the member-states, but built to enable them to revive and peacefully interact with each other. For nearly 60 years the EU has been slowly and generally benignly blunting former national antagonisms, institutionalising co-operation and encouraging member-state interaction and learning through billions of daily local micro-interactions.

It has progressed through co-operative innovation, adjustment and evolution, not through centralised control or a rigid plan.

Given the EU's complex nature, Brussels is an excellent non-traditional 'capital'. Its position in a small, centrally located, non-threatening member-state that couldn't begin to impose political, economic and/or cultural hegemony over the rest of Europe makes it an ideal 'non-capital' city. To prove this point one merely needs to consider the historical landmines of moving the EU capital to Berlin.

Happily, the end of the proposal for a 'capital for Europe' was a healthy complex outcome that included a sprinkling of initiatives for promoting Brussels and a few small development initiatives. Major pedestrianisation and humanisation issues (putting the noisy and smelly Rue Belliard underground) were sidestepped due to cost, but thankfully fantasies of a traditional capital for the EU were buried as well.

Using a complexity cascade to visualise the development of the EU, Scandinavia and Brussels

As we have argued throughout this book, one of the key problems of the orderly perspective is the desire to find and impose a final model/endpoint on a given social institution or structure that would permanently lock it into place. From the nineteenth century writings of Karl Marx to the recent work of Francis Fukuyama and the US neo-conservatives, history was seen as moving through key historical stages and towards an ultimate direction. For Marx it was the movement from a divided feudal society (multiple classes – peasants, artisans, merchants, landed gentry, priesthood, aristocracy), to a binary division within capitalist society (two classes – workers *versus* capitalists) to a unified socialist/communist society (all unified in one class). For Fukuyama and others (Samuel Huntington in particular), history was going through a similar transition from a divided feudal/non-democratic society (before the rise of modern democracy), to a binary division within global society (Cold War struggle between capitalist, liberal democratic systems and communist, anti-democratic systems) and a unified capitalist liberal democratic world system. At the EU level, the underlying idea for orderly pro-EU actors was that the EU would eventually unify the various member-states into a coherent whole. From this perspective, once the endpoint is reached the system stops evolving and remains stable (reaching what could be called its final equilibrium point). History comes to an end and time no longer matters. Visually, this perspective looks something like Figure 6.3.

As we saw in the earlier chapters, complex systems do not come to a final resting point nor are they totally random. As described in Chapter 3, one way of visualising a complex system is through a complexity cascade. In a complexity cascade, the system (and time) never stops while it moves through a series of changes in a punctuated equilibrium fashion. Gateway events create new opportunities for change. Following this, frozen accidents begin to occur that create a growing number of regularities in the system. The system stays stable for a time until a

124 *The international arena*

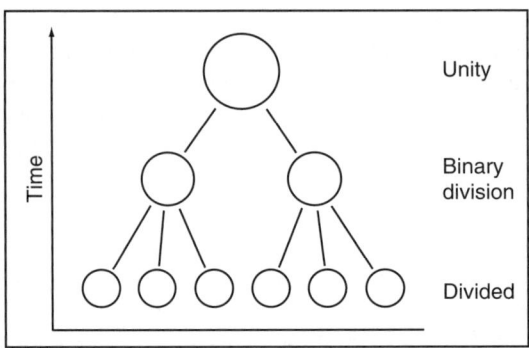

Figure 6.3 The traditional model of unity and order.

new gateway event creates new opportunities for change. Given this, what would a complexity cascade of the European Union look like? Figure 6.4 attempts to answer this.

In this case our cascade begins with the gateway events of the Treaties of Paris (1950) creating the European Coal and Steel Community and the Treaty of Rome (1957) creating the European Economic Community. The two events set the core framework for the future EU (including aspects like economic cooperation and interaction, West European orientation, democratic institutions, etc.). There were many elements that were outside of this core framework (pure market institutions, members outside of Europe, non-democratic regimes) and these would fall into the 'A' zone on the picture. However, there was also a great deal of openness that these treaties implied (exact members, economic rules and developments, political cooperation – hence the open 'V' pattern). As time moved on, developments and frozen accidents (European economic recovery, European institutional entrenchment, British opposition to membership, support from the US, etc.) created more and more regularities within the system. Eventually a new gateway event occurred that radically transformed the former system with new opportunities and challenges. In this case, the Single European Act and Single Market White Paper signal the radical deepening and expansion of EU policy in the late 1980s. This led to new frozen accidents (expansion of economic integration, limitations in social policy, new forms of European economic interaction), that created new regularities (new power balances within the EU institutions, new member-state demands and policy developments). Again, a new gateway event occurred in 1989–93 with the end of the Cold War and creation of the Maastricht Treaty. Subsequently new frozen accidents (new East European applicants, the reintegration of East Germany, new policy demands) led to new regularities (new voting procedures for policy decision-making, new powers for the European Parliament, new demands for EU foreign and security policy) and so on. The latest gateway event, the expansion of EU membership to the east (now 27 member-states – a massive change from the original 6) and creation of European Monetary

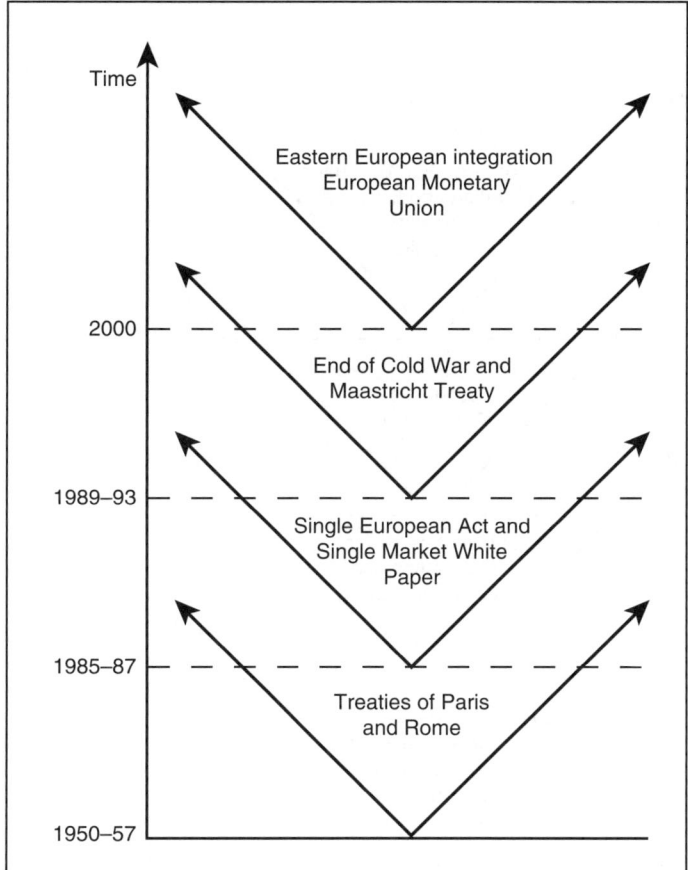

Figure 6.4 European Union complexity cascade.

Union (creating a common currency) has once again opened up a whole new range of opportunities and threats. Frozen accidents have already occurred (the success of the euro and struggles of the dollar) and new regularities are starting to take shape (new voting patterns in the European Parliament and Council). Who knows what the next major gateway event will be. Following this pattern, Figure 6.5 shows that similar cascades can easily be constructed for the development of the Scandinavian social democratic societies.

In this case, during the 1930s all three Scandinavian countries experienced an expansion in trade union activities and the creation of their first social democratic governments. This set up frozen accidents (key trade union and social democratic actors gained policy and governmental experience) that created new regularities (welfare state policies, rights for workers) that remained stable despite the interruption of WWII. Following the war, all three countries experienced a new

126 *The international arena*

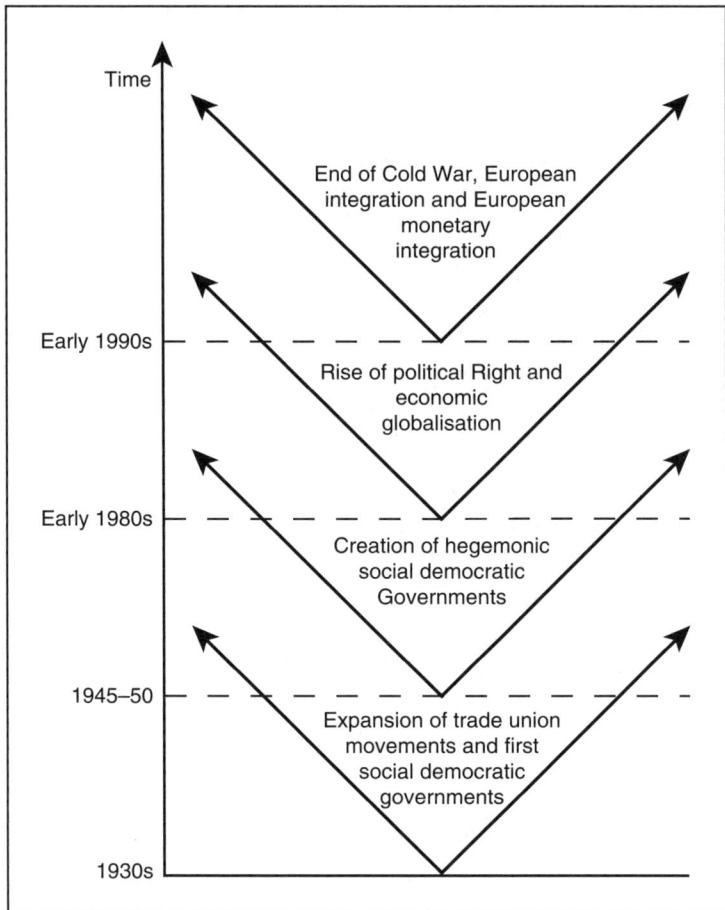

Figure 6.5 Scandinavian social democratic complexity cascade.

gateway event through the election of dominant social democratic governments that lasted through the 1950s and 1960s. Again this led to new frozen accidents (growing service sector jobs, increased female participation in the labour market), that created new regularities (extensive universalistic welfare entitlements, new professional class). Later, the early 1980s saw the rise of the new Right and impact of globalisation creating new frozen accidents (fall of social democratic parties, growing pressure on welfare state budgets) leading to new regularities (restructured welfare provision, new types of labour rights). Finally, the end of the Cold War and intensified European integration mark the latest gateway event leading to new frozen accidents (joining or not joining the EU) and new regularities (new relationships to EU members).

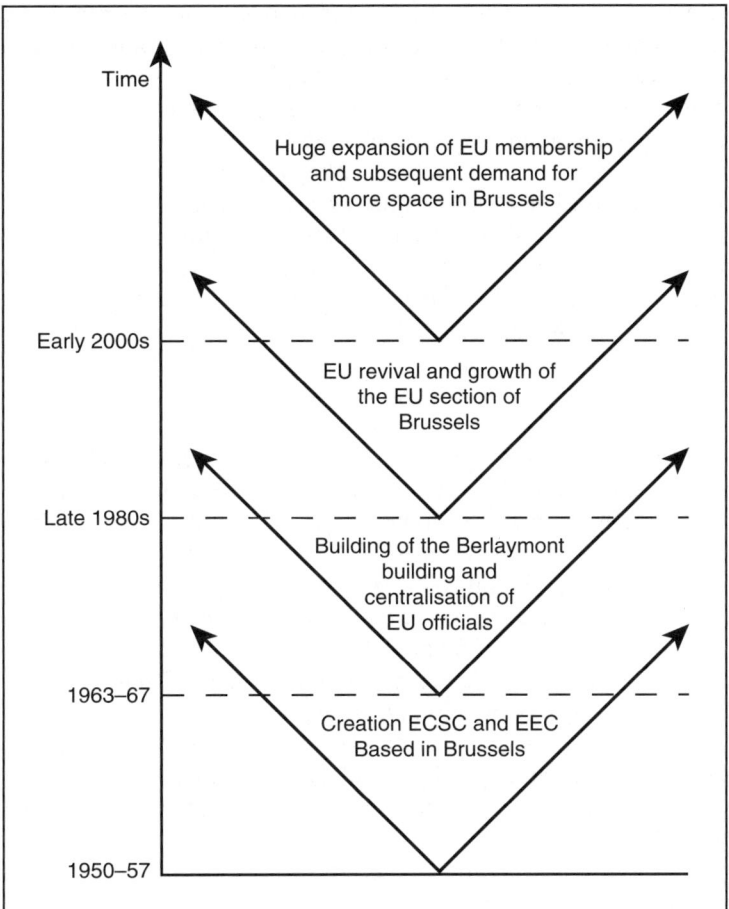

Figure 6.6 Brussels as a capital for Europe complexity cascade.

In this final case, presented in Figure 6.6, the gateway event of the creation of the European Coal and Steel Community (ECSC) and European Economic Community (EEC) created a new framework for the development of Brussels that led to new frozen accidents (new bureaucrats being stationed in Brussels) and regularities (need for new buildings to house them). This led to a later gateway event, the building of the Berlaymont building in central Brussels to centralise the growing numbers of European bureaucrats. Unsurprisingly, this led to new frozen accidents (growing impact of bureaucrats on the city) and new regularities (new roads to accommodate greater traffic in the area). Later, the revival of the EU in the late 1980s led to a huge growth in the number of interest groups and lobbyists that moved to Brussels (frozen accidents) that led to the recognition that part of the city was now the EU section (regularity). Moreover, most recently with the expansion in EU membership and construction of the new Parliament building, even more EU actors have poured into the city (frozen accidents), making a large part of the centre completely dominated by EU actors (new regularities) and leading some to demand that Brussels should be the new capital of Europe!

The key point to recognise with all of these examples is that there is no final endpoint merely a continuing series of emergent developments to complex systems. Earlier, developments delimit or set the framework for the future. Hence, as many have said before one cannot escape the past. However, new gateway events, frozen accidents and regularities are constantly opening up new possibilities and constraints. Just as complex systems are a mix of order and disorder, the present is a mix of the framework of the past interacting with the possibilities and choices of the future.

Conclusion

What do these case studies tell us about the nature of the international system and the academic attempts to understand it? Fundamentally, they reconfirm the 'commonsensical' notion that the international system, from the biggest conflicts to the minor actors and details, is a remarkably complex, evolving, emergent and adaptive system. Complexity allows us to recognise the true nature of the system, gives us some new tools to explore the system and enables us to see the limits of traditional 'orderly'-based interpretations. Another amazing aspect of a complexity perspective is that it enables one to see similar processes and dynamics at all levels of the system, from globalisation and Europeanisation to Scandinavian political economy and the politics of Brussels. There are obviously a whole host of other relevant examples and complexity inspired thinkers are just beginning to explore them. We will touch on some other pertinent aspects of the international realm in later chapters.

7 Development[1]

> We must embark on a bold new program for making the benefits of our scientific advances and industrial progress available for the improvement and growth of underdeveloped areas ... Greater production is the key to prosperity and peace. And the key to greater production is a wider and more vigorous application of modern scientific and technological knowledge.
>
> (President Harry Truman, 1949 inaugural address)

The underlying political economic tenets on which today's development policies and actions are founded were spelt out in Truman's groundbreaking 1949 inaugural address. From this perspective, development was seen predominantly as economic growth to be achieved through actions similar to those presumed to have been adopted in the past by the richer countries. Based on the orderly pillars of neo-liberal economics, the idea of planned 'scientific' development proved to be extremely attractive and resilient. The task was envisioned as a finite project with clear beginnings and ends that would be achieved in a pre-planned process and according to laws of universal applicability. Leading powers and international bodies such as the World Bank, guided by the successful experience of reviving the Japanese and European economies after WWII, arrogated for themselves the primary role of uplifting the fortunes of less fortunate nations. These beliefs enshrined an implied fundamental principle that was not explicitly articulated: economic development would in time lead to human development. This view was based on what seemed to be a sensible argument that growth would make available the financial resources needed to improve health, education and other basic services. Economic capital was substantially divorced in that way from human capital.

Within that overarching framework, the World Bank adopted and then discarded a succession of economic growth theories based almost exclusively on selected data from Western economies, such as the Harrod–Domar model and then the Solow model. Throughout, there was emphasis on loans and aid to fill a ubiquitous 'financing gap' (Easterly 2001). The focus of this chapter, therefore, is to explore the potential that complexity offers to the theory and practice of development, examine the substantial, and necessary, upheaval associated with a move from an orderly to complexity framework for development, and describe a selection of policies and actions to turn this shift into reality.

The case for radical change in development practice

The onus to prove the need for any change rests squarely on those proposing such action. We say 'any' change because we feel it is necessary to establish beyond doubt that the development project which has been going on since the end of the Second World War has failed comprehensively. We will deal briefly with this before we turn our attention to the nature of the change needed to produce better outcomes.

There are now a multitude of works that have exposed the failings of development policy and theory (Caufield 1996; Hoogvelt 2001; Rich 1994; Stiglitz 1998). Of particular interest is the work of William Easterly (2001), a former World Bank economist, who described the corrosive influence successive economic growth theories exercised on actions undertaken over several decades, even when their failure was all too obvious. More recently, Ha-Joon Chang, a Cambridge University economist who specialises in development studies and won the Gunnar Myrdal prize for *Kicking Away the Ladder: Development Strategy in Historical Perspective* (2002) and later wrote *Bad Samaritans* (2008) and Angus Maddison, a renowned economic historian, in *Contours of the World Economy* (2007), attacked the prevailing development wisdom from their differing perspectives. They argued comprehensively that today's rich nations did not adopt slavishly orderly and immutable recipes on their way to the top. On the contrary, their eclectic progress followed diverse and uncertain routes. The only notable attempt at a rigid orderly model was that adopted by the Soviet Union which ended in utter failure. The United Nations Development Programme (UNDP), as opposed to the IMF/World Bank, has been tireless in pointing out, through the annual Human Development Reports and other publications, that economic development can not be dissociated from human development.

Despite all of this work, the increasingly obvious growth of global inequality (see Figure 7.1) and the blatant failure of large scale financial loans to stimulate economic growth in the poorer countries (Easterly 2001; George 1994), the search continues for orderly strategies and actions of universal applicability that might offer better results. Fleeting visits by missions from the IMF and the World Bank and dogmatic adjustment programmes comprised of a few steps to suit all nations, to cite only two examples from many, have been shown to produce haphazard results and regular failure.

Without labouring the subject any further it is possible to argue that there is an urgent need to radically rethink development. The more substantive question is what to do next? What change is required in the way development is envisioned and pursued that would promise more reliable outcomes? To do this, we need to first understand why the view of development became so absorbed with order.

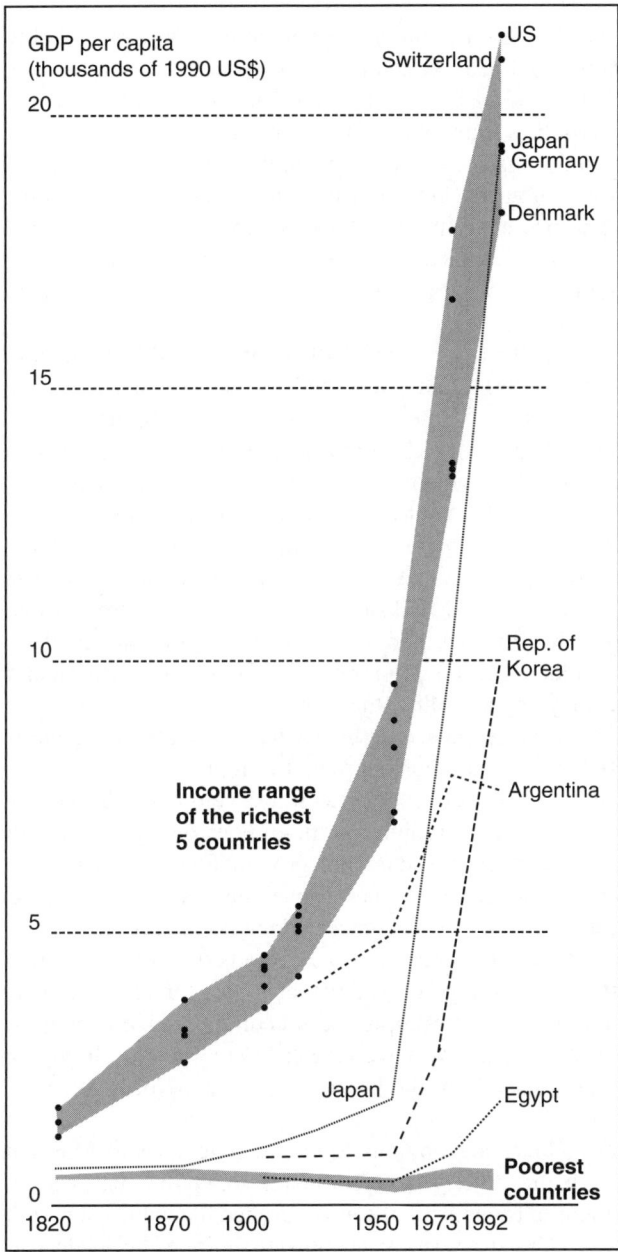

Figure 7.1 United Nations Development Programme, *Human Development Report*, (1999: 38).

Lopsided view of the way nations evolve

In many ways, the post-WWII vision of orderly development was a remarkably attractive one. Given enough political will, financial support and scientific (i.e. orderly) expertise any nation could be transformed. This vision mirrored the dominance of the orderly policy framework, fit in with a Cold War vision of developing allies in the fight against communism and recognised the end of colonial empires and growth of newly independent states. It was also conducive to the belief in the role and power of highly educated experts who had studied 'development', knew how it functioned and could apply it anywhere in the world. Unfortunately, these beliefs led to a range of maladaptive assumptions and strategies.

For example, in order to establish that any and all countries could be radically accelerated at any time, the orderly framework had to demand that everyone ignore the basic economic history of most of today's leading economies. In general, the rich have been rich for a very long time and their so-called miraculous economic growth was achieved in the main through slow, modest, and laborious growth; typically of just over one per cent compounded over lengthy periods – in some cases stretching over several centuries (Chandler et al. 1997; Maddison 1982). Similarly, the orderly approach had to ignore the fact that success was accomplished through gateway events; industrial revolution, colonisation and exploitation, discovery of new territories, and so on, that were now unavailable to those at the bottom of the ladder. This does not mean that poorer countries cannot develop, but implies that they cannot follow the same pathway that presumably wealthy countries followed because the economic, social and political structures within which they are evolving are so fundamentally different.

The use of the word 'evolve' is itself at odds with a development orthodoxy conceived mainly within an orderly paradigm; a one-dimensional process with finite beginnings and ends, focused on economic growth founded on universal recipes. The misleading distinction between 'developed' and 'developing' countries lies at the heart of an orderly vision that does not stand up to even cursory scrutiny. As Rihani (2002) pointed out, no one could credibly suggest that Europe or the US have stopped developing. And yet, evidence for such beliefs is unmistakeable. Rostow (1960), for instance, reduced economic development to five distinct stages which was likened by Toye (1987: 11) to a move 'through a series of stages derived essentially from the history of Europe, North America, and Japan'.

As we have seen earlier, there are obvious hints of Hegelian and Marxian 'historical inevitabilities' in the orderly developmental viewpoint: nations are dotted on a common escalator of progress; some have reached the final destination and others are on the way. The stragglers need only to imitate the leaders to achieve ultimate success. However, historic evidence (such as Maddison (2006, 2007) describing the evolving mix of realist and liberal strategies; ranging from the ruthless use of force to more liberal attitudes founded on international trade and cooperation) stretching over several centuries of the variegated ways in which today's rich countries rose to the top of the economic league has to be ignored by the orderly paradigm.

Finally, the orderly framework demands a reliance on top-down management styles and the growth of a substantial 'development industry' (Sen 2000) directed from above through the United Nations, the World Bank and IMF, world leaders, and specialist governmental agencies. Local participation and appropriate technologies were given scant attention despite their obvious success in the few instances where they were tried. Grand policies and major infrastructure projects were preferred in the belief that growth could be initiated and speeded up at will; it simply required determination, financial resources and modern infrastructure. In the event, and some would say as intended, these projects offered rich pickings to the richer countries and negligible benefits to the recipient nations (Caufield 1996).

Why do the successful succeed? Self-organised complexity

Kauffman (1993: 173) pointed out that complexity has valid applications in 'living systems, organisations, communities and coevolving ecosystems'. Others, such as Beinhocker (2006) in economics and Blackman (2006) in urban development identified parallels between self-organised complexity and social phenomena. Rihani (2002) advanced similar arguments but this time directly in relation to development. A few examples drawn from the 'developed' countries will help in outlining the argument for viewing wealthy 'Western' countries as cases of successful self-organising complexity.

First, these countries present a stable general pattern typified by basic democratic and market structures. Within the framework of this global model there is massive variety, which gave them the sustainability to overcome numerous challenges. They were prepared at times to adopt other political economic strategies based on state-centred realism for instance or radical variations of liberalism, as was the case in Keynesian macroeconomics. This pragmatic setup is remarkably reminiscent of the healthy variety provided by *states* within one *stable attractor* in self-organised complexity. One can easily see it at the end of the first decade in the twenty-first century in the abandoning of neo-liberal economic rules in response to the credit crisis (nationalising of much of the banking sector, huge deficit spending and constraints on international financial movements, etc.).

Second, governments of richer nations appreciate the value of 'social capital' represented by the diverse activities of individuals and groups. Samuelson and Nordhaus (1995: 299–304) reported that the US government's public spending patterns changed little under different administrations. In essence, high spending (relative to most poorer countries) on law enforcement, nutrition, health, education and income protection, is dictated by the wish to endow citizens with the freedom and ability to pursue their varied interests and in that way to contribute to the general wellbeing of their community. The recent work of the esteemed political scientist Robert Putnam, *Bowling Alone*, (2001) clearly demonstrates the serious concern that nations should have for their social capital. The importance of connections between individuals and the ability to connect is well established in complexity.

Third and associated with the above point, development in the richer countries stemmed largely from uncoordinated efforts by individuals and groups concerned with their businesses and individual pursuits working within a relatively stable social framework. The uncoordinated and unplanned emergence of the industrial revolution in Britain, driven by inventions such as Hargreaves' spinning jenny, Arkwright's use of waterpower, Watt's steam engine and Crompton's spinning-mule, was not a planned project. Bill H. Gates, of Microsoft fame, is an outstanding example from the twentieth century of this self-driven endeavour that was not preordained by an external authority. The spreading ripples of his impact on almost every aspect of daily life are all too obvious.

Fourth, as argued for instance by Ha-Joon Chang (2003, 2007), present-day richer countries followed an evolutionary and eclectic path characterized by many twists and turns over *very* long periods. They did not follow a specific set of prescriptions either individually or as a group. This continues to be the case. Their evolution is implicitly accepted as being slow, open-ended, and unpredictable. The process has also exhibited clear signs of *punctuated equilibrium*; long periods of apparent tranquility interspersed by brief episodes of radical transformations, and *gateway events*; such as the invention of the steam engine and computers. Yet again, the parallels with complexity are evident.

Fifth, today's developed countries accumulated complexity and acquired 'depth' over several centuries. They are becoming richer, and the gap between them and poorer nations is widening; a feature that attracts much comment and concern. But that is precisely what one would expect if the development of nations accorded with the standard behaviour of complex adaptive systems: average complexity increases and the highest complexity stands to gain the greatest growth. Likewise, complex adaptive systems are much more able to absorb shocks than less complex systems and they demonstrate better recovery after setbacks. The example of the impact of the fall in the price of copper had on Zambia (whose only major export is copper) is well known, but the impressive recovery of the European and Japanese economies after World War Two is a more dramatic illustration of the sustainability of more complex social, political, and economic organisations. By contrast, observers often suggest that Iraq will need several decades to recover from the travails that ensued from the 2003 war when all its systems, and not just its physical infrastructure, simply disintegrated in a few months.

Why do the unsuccessful stagnate?

Barriers to development

Countries leading the field selected, through lengthy trial and error, practices that optimised their performance. There was no need for them to know anything about complexity or complex systems. Conditions in the less advanced countries are in general the exact opposite of those required for these nations to assume a stable but evolving pattern of self-organised complexity. For example, Rihani (2002) argued that the internal elements of an evolving nation are interacting human beings, as individuals or groups. Too few interactions result in a state of stultifying order, while too many could lead to chaos. The concept of balance and range of outcomes tool discussed in Chapter 3 typifies this point. The two extreme states are reminiscent of rigid order in Iraq under Saddam and then the chaos of the post-Saddam era. Both conditions lead to a developmental dead-end.

Basically, the layer of self-organised complexity that lies at the so-called 'edge of chaos' could only emerge if individuals were *free* to interact and *capable* of interacting, and if their interactions were facilitated by *appropriate rules* that command popular support. Absence of these attributes could, and often does, result in lack of sustainable progress or even regression. This is not a new idea. Adam Smith advanced his recipe for economic success more than two centuries ago: allow people to barter and trade freely and leave it to the 'invisible hand' to produce self-generated order in the marketplace within an open and stable framework of national and international rules. Complexity simply provides the technical tools to explain why this is so.

Few of the developing countries meet the *freedom* criterion. State repression, against whole populations or sectors defined by gender, religion or ethnic background, is widespread. The pattern of control by privileged and ruthless elites that set out to stifle diversity of independent action is all too common. However, it is essential to underline the damaging, and self-evident, association between repression, corruption, militarism and conflict. The significance of these impediments to development has been well understood for quite some time. UNICEF declared in the *Progress of Nations 1997* that 'violence against women and girls is the most pervasive violation of human rights in the world today ... its impact on development profound'. Ten years later UNICEF went back to the same topic in *The State of the World's Children* which was given the title *Women and Children: The Double Dividend of Gender Equality* (2007). Similarly, Collier argued, 'democratisation is worth around half a century of income growth in terms of its contribution to peace' (Collier 1998: 18).

Poorer countries suffer from a massive freedom deficit in an even more tangible way: they are beset by external pressures from a number of sources, be it quarrelsome neighbours, world-leading powers, international organisations, and/or the globalised world economy. As an extreme example, John Perkins (2004) described, in *Confessions of an Economic Hit Man*, the punitive measures adopted by more powerful countries to impose inappropriate actions on nations in urgent need for development. Misunderstandings about the real nature of the development process fit in rather well with self-interest to make the task of moving to a more effective paradigm even more difficult.

The factors that affect individuals' *capability* to interact are equally clear. The main culprits are malnutrition, disease and illiteracy. Again there is nothing new about this. A Report of the Sanitary Commission of Massachusetts was presented to the state legislature in 1850 that remains to this day the accepted wisdom on how to combat ill health at low cost (Evans *et al.* 1981). WHO and

UNICEF in conjunction with over 130 countries launched a campaign in 1978 to achieve 'Health for All by the Year 2000'. The outcome from this and other efforts have been disappointing. And yet the consequences of failure are also common knowledge. UNICEF (1998: 9) referred to malnutrition as the 'silent emergency'. There is a close and well-understood link between malnutrition and disease. Shortage of vitamin A, iodine and iron, and poor standards of sanitation and water supply are just as significant as food shortage. Remedial measures, as in the case of vitamin A, amount to no more than a few pence per person per year; see UNICEF (1995: 18). UNICEF again reverted to the same topics in *The State of the World's Children 2006*. The report pointed out that some 1.9 billion children live in developing countries and almost half live in poverty. And 121 million primary school-age children are out of school, mostly girls. Is it any surprise that economic growth is illusive under these conditions? One does not require complexity to recognise this feature but complexity helps to explain why development is a non-starter unless human development is addressed as an integral part of the process.

Apart from being free and able to interact, the actions of groups and individuals have to be facilitated by simple rules to enable a nation to achieve a state of evolving self-organisation. Yet again, many poorer countries are far from meeting these basic conditions. De Soto (2000: 15–20, 63), for instance, described the hundreds of bureaucratic steps, and the years, required in Peru, the Philippines, Egypt and Haiti to start a small business or build a modest dwelling. Similarly, corruption is a major problem, creating an unstable system where basic laws are continually flouted. The World Bank realised years ago that 'corruption ... is negatively correlated with both investment and growth' (World Bank 1997). A complexity view of development simply provides a clearer framework for why these 'common sense' ideas are correct.

Once again, there is nothing new about these imperatives. The Brundtland Report (World Commission on Environment and Development 1987) recommended specific changes in attitudes and practices needed to achieve better performance. Ideas advocated included sustainable development, self-help, focus on basic needs, respect for indigenous knowledge and local coping strategies, and emphasis on institutional improvements. These and other innovations were arrived at intuitively as sensible improvements or obvious responses to policy failures. However, they lacked an overarching theoretical framework to bring them all together and defend them against the criticisms of being 'soft' or unscientific. Complexity can provide just such a foundation.

A more realistic view of development

What does all this imply? As in the natural world, the evolution of nations proceeds as a cyclical activity that has three indivisible components: survival, adaptation and learning. Survival requires a discernable stable structure; learning means the build up and application of relevant knowledge, while adaptation describes change that enhances performance and promotes survival. Success, therefore, relies on striking a balance between: malleability, but not so much that the slightest shock would destroy the structure; and stability, but not so much as to prevent adaptation (Kauffman 1996: 73). The breakup of Iraq after 2003 into warring factions is an example that illustrates the point unambiguously.

A nation behaving as a healthy complex adaptive system acquires self-organised stable patterns through numerous local and external interactions involving all elements of the system, as individuals, groups, governments, and so on. The interactions have to be regulated by simple rules that command general acceptance, as haphazard interactions produce chaos without a pattern, and rigidly controlled interactions result in an unchanging pattern that could not evolve (Stacey 2000: 281). The cyclical process of survival, adaptation and learning, similar to that found in complex systems in the natural world, requires *time* (Dawkins 1996). Hence, from a complexity perspective, development is a slow, tortuous and long-term process that can rarely be rushed. This also implies it is neither orderly nor predictable.

In brief, complexity views development as an uncertain, open-ended, and long-term process driven by a large number of interactions that generate self-organised stable patterns capable of adaptation. Furthermore, a nation is part of a nested structure of complex adaptive systems that range from other neighbouring evolving nations to the global system as a whole. External actors might be able to influence changes in the system. However, the impact of their actions will be uncertain and limited at best. The most effective actions happen at the lowest possible level. An approach often espoused by the rich world internally, as illustrated by the EU's concept of 'subsidiarity' (pushing most decisions down to the lowest level) or of demands for 'states rights' in the US (giving more power to individual states), but often ignored by international organisations when imposing policies on the rest of the world.

One key feature associated with the above concerns compulsion and the use of force. A doctrine has acquired credence that commends the use of military force, or the exploitation of natural or man-made disasters, as an effective and fast way to introduce democracy, liberalism, and development to nations and communities that are seen to be deficient in one or more of these attributes (Klein 2007). In general, complex adaptive systems do not respond well to brute force. Put more precisely, to change the behaviour of a complex adaptive system quickly requires massive force which has to be maintained to keep the system on its new path. Such systems have inbuilt ability to test and, when necessary, resist external influences. This is sometimes referred to as the 'culture' of the system, and is part of the mechanisms by which the system protects itself against possibly harmful intrusion. As demonstrated in our chapter on health, this is a common aspect of most social systems and institutions. A complexity framework for development strongly suggests that internal or external imposition of change, no matter how desirable and well-intentioned, is destined to be a costly project that often ends in failure. As we will explore in a later chapter, the constant recasting of the US mission objectives for Iraq since 2003 is a prime example that should serve as a lesson for the future.

138 *Development*

From *x–y* to fitness landscape thinking

As presently conceived, the development project seeks to guide a nation, or community, 'in need of development', along a predetermined path, with supposedly clear signposts that have been charted by earlier pioneers, to a specified end-state of 'development'. Ideally, from a traditional framework, this path could be mapped onto a common *x–y* graph as shown in Figure 7.2.

The country to be developed would be positioned on a certain level on the development ladder or escalator, such as point A or B, and appropriate policies would be applied until it reached the stable point of the fully developed countries, point C. Once it was there, having reached 'maturity', it could continue on its own way for the foreseeable future to point D. The implications for the country to be developed are simple: do as you are told by the international wealthy elites and copy what they have done and everything will work out for the best.

As demonstrated earlier in the chapter, life is not that simple. Basically, there are many pathways to development as seen by the experience of today's rich countries. Even more disconcerting, each of these countries adopted on their way to the top several pathways as varying conditions dictated. They had to be pragmatic; twists and turns of circumstances were largely unpredictable. In short, successful countries took a variety of directions/policies/institutions/cultures and created, over lengthy periods of time, distinctive outcomes that can be generally packaged as being 'developed'. A linear *x–y* graph could hardly be used to present an illustration of that concept of development. Visually and metaphorically, the next step away from traditional thinking is to visualise development occurring on a three-dimensional fitness landscape, as is shown in Figures 7.3 and 7.4.[2]

In this simple fitness landscape, we combined three factors: time 1971–2006 (*z* axis), annual growth rates (reported by the United Nations National Accounts) on the *y* axis and a range of how disorderly – orderly the society was on the *x* axis.

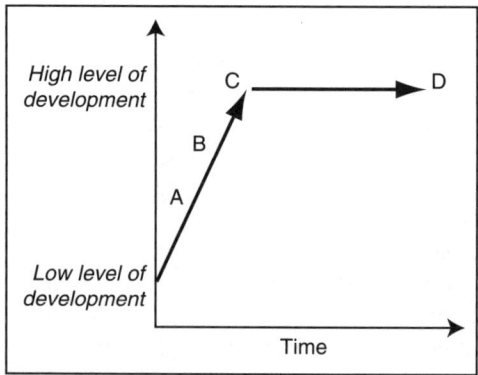

Figure 7.2 An *x–y* view of development.

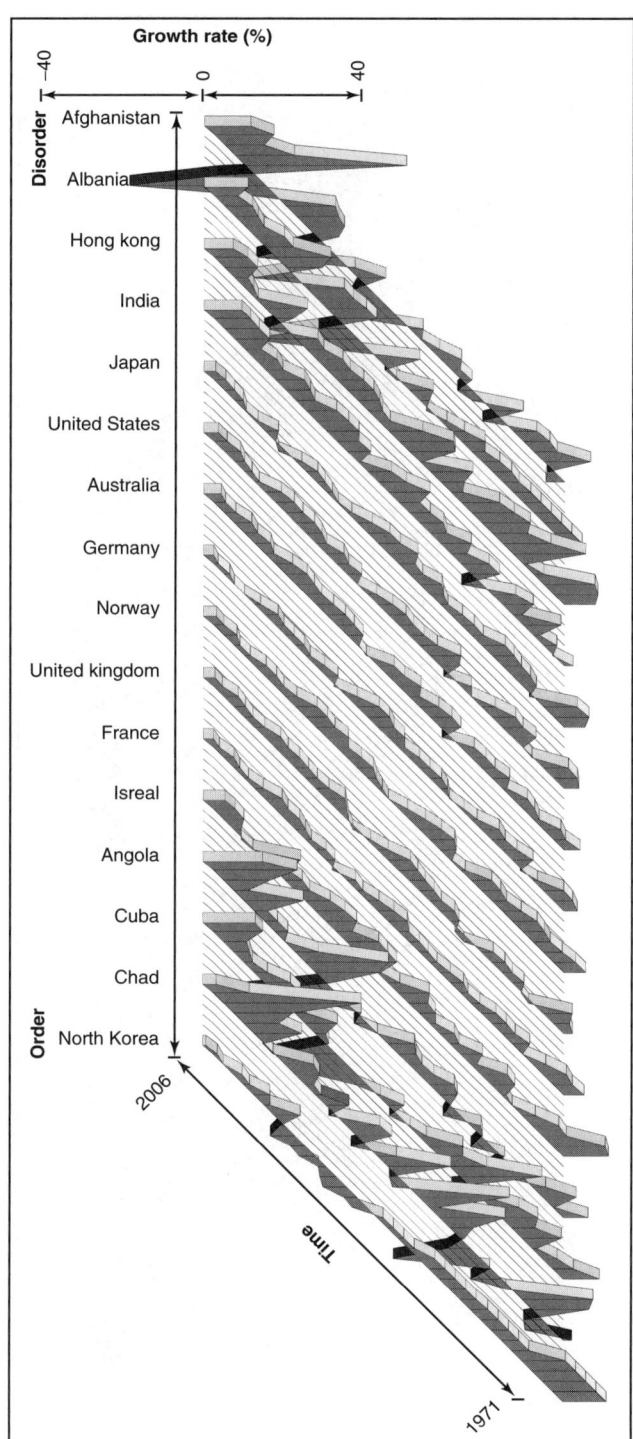

Figure 7.3 A fitness landscape view of development.

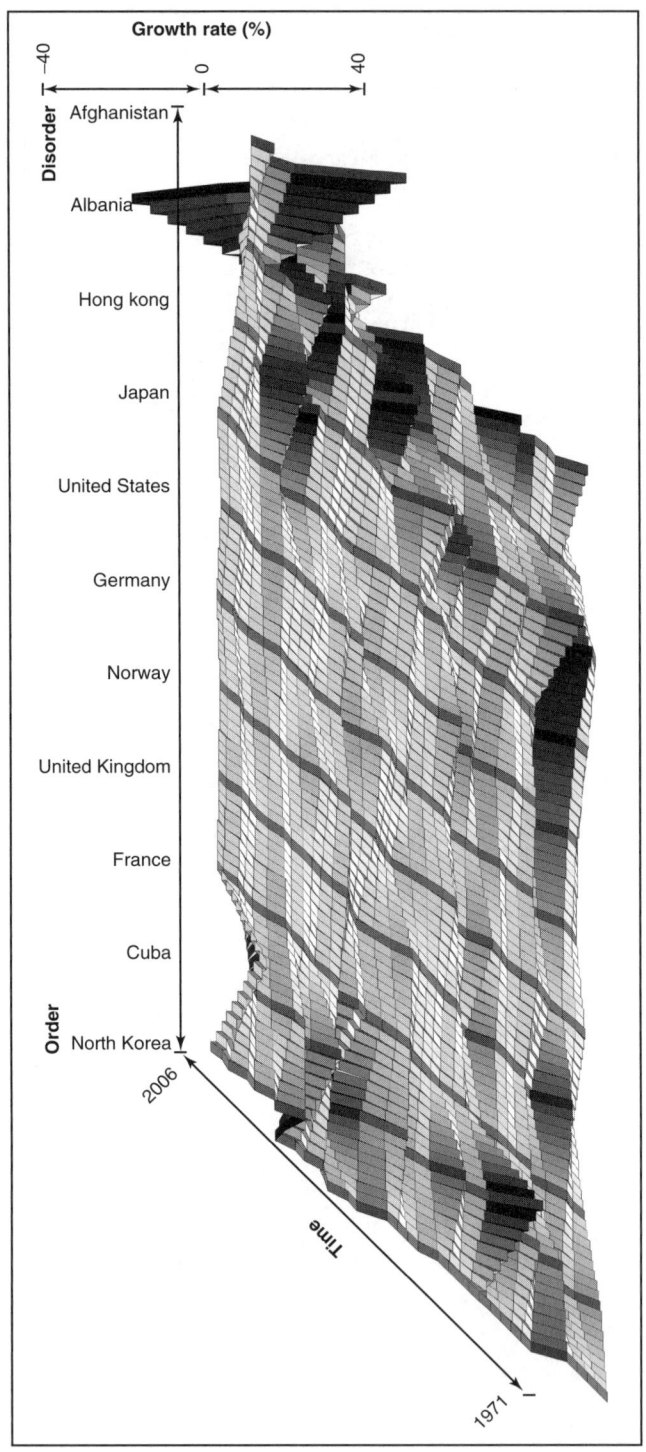

Figure 7.4 Another fitness landscape view of development.

For the disorder – order scale we used the UN National Accounts and looked at state expenditure as a percentage of gross national product (a very general indication of how much the state controls the economy). The more state control, the more order. The less state control, the less order. Obviously, this is only a very rough indication of how much a society is held in an orderly state or caught in a disorderly conundrum. Nevertheless, some very revealing patterns emerged.

For the advanced industrial countries:

- Unsurprisingly, they tended to be bunched towards the middle, exhibiting a mix of state-market/order-disorder (avoiding extremes of order or disorder).
- More interestingly, despite being in the middle there is a remarkably broad range of market-state or disorder-order variety. For example, France (on average) has nearly twice as much state dominance of the economy as Japan and almost three times as much as Hong Kong.
- Moreover, for most cases, the key to success is small-scale improvements over long periods of time. As the data shows, with some famous exceptions, there is no 'rush for growth'.

For the less economically successful countries:

- Being too orderly or disorderly can be equally bad. On the orderly extreme, North Korea reported reasonable growth rates until the mid-1970s and then went into a steady decline as it became more and more rigid and less and less capable of successfully interacting with the increasingly globalised world.
- At the other end of the spectrum in a case like Afghanistan where there is a lengthy history of civil division and strife and almost constant war since the late 1970s, growth rates fluctuate wildly with the constantly changing security situation.
- Unsurprisingly, both of these cases score very poorly on Transparency International's rating of corruption indicators (see http://www.transparency.org/)
- Finally, for a case like Iraq (not included on our landscape but discussed in the next chapter) which went through a period of stifling order under Saddam Hussain and then was pushed into a state of chaos by the second Gulf War and occupation, the prospects are obviously bleak.

What are the general implications of this re-imaging of development from a complexity perspective? Development is revealed as a process of exploration with a desired general direction in mind. There are some rules to follow; create a stable institutional framework, encourage decentralised local interactions, avoid civil strife and a stifling state structure, and so on. However, there is no perfect pattern for exactly how a country should stay within these boundaries. Hence, detailed prescriptions of what should be done in the long term are ineffective or counterproductive. At best, external actors should encourage developing states to create stable frameworks for allowing their citizens to interact freely and effectively, to learn, adapt and survive, and hence to improve their human and economic prospects under a wider range of conditions. This, with luck, could turn a 'non-developing' country into a 'developing' one.

Why changing the framework of development is so slow?

Given all of this, it is not surprising that long-standing calls for a change in direction have come from many quarters including the World Bank itself. The proceedings of the tenth Annual World Bank Conference on Development Economics held in Washington DC in April 1998, for instance, gave a clear indication of dissatisfaction with past performance. Another authoritative statement on the subject appeared in late 2000 when the World Bank published its *World Development Report*. The ensuing vigorous debate led to the resignation of the main author of the report, Ravi Kanbur, and prompted one commentator to ask 'why it took 50 years to reach this point' (Brown 2001: 13, 157).

What is surprising, however, is that despite these and later expressions of discontent with current practices the development industry seems to be no further on the way to adopting more effective strategies and actions. We contend that this muddled state will continue for as long as development is misdiagnosed, wrongly, as a finite linear process that has easily identified causal relationships encapsulated in simple laws of universal applicability. A fundamental misunderstanding of the nature of development was bound to lead to the selection of ineffective measures. Recognising development as a complex adaptive process would overcome these conceptual and practical difficulties. A change along these lines would result in a strategic approach to development that addresses the factors that determine whether the process functions properly or otherwise.

For this to happen, a number of difficult transformations will have to be accepted. The first and most significant transformation concerns the elevation of people to centre stage in the development process. Rather than reasonableness or correctness, this shift in the balance of power is an inevitable outcome from the recognition that local interactions are the driving force behind the emergence of self-organised stable patterns that are capable of effective adaptation. Ultimately, local actions determine whether a nation develops or stagnates.

The second transformation flows naturally from the first. If local interactions are paramount then people, as individuals and groups, must be able and free to interact, otherwise nothing happens. In essence, *human* development; primarily a focus on basic needs and human rights, would have to be given top priority over all else, including economic development. Rihani (2002) and Rihani and Geyer (2001) traced the rationale behind this transformation and its wider policy implications.

The third transformation again follows from previous remarks. Interactions between the internal elements of complex adaptive systems must proceed in accordance with simple rules if self-organised and stable, but evolving, patterns were to emerge instead of wasteful chaos or stultifying order. Consequently, institutional and democratic reforms within a complexity view of development become urgent imperatives as opposed to optional extras to be implemented 'when conditions permit'.

The above transformations carry a vital message: the development process could not be initiated and sustained without a shift in focus onto local issues concerned with basic needs, institutional reforms and more enlightened democratic practices. However, one further transformation is required to complete the picture. Success within the complexity framework depends on substantial relaxation of present-day local and global rigidities (Rihani 2002). Efficient exploration of the nation's fitness landscape relies heavily on copious diversity, readiness to experiment and make mistakes, and a pragmatic outlook that shuns inflexibility in all matters.

A number of scholars have discussed the healthy tension that must exist between harmonising and universalising factors on the one hand, such as science, technology and shared visions and aspirations, and the diversifying factors that enable communities, organisations and other complex adaptive systems to 'display the internal capacity to change spontaneously' (Stacey 2000: 386–409). A glance at today's non-developing countries would show that they are generally not in a position to enjoy the benefits of this duopoly. Internally, they are riddled with political, social and cultural inhibitors to diversity. Externally, they are at the mercy of a global system that insists on compliance with universal norms that perceive any deviation as a challenge that should be nipped in the bud. The dominance of the neo-liberal economic model, at least up to the so-called credit crunch of 2008–9, has been possibly the most significant obstacle to healthy diversity and, hence, development.

China's sprint to the top of the development league reveals the progress that could be made by a country when it is able to resist dictates from others. As was the case with Britain and the US a century or two ago, China experimented with all three leading schools of thought in the political economy – Marxism, realism, and liberalism – and reaped the benefits of that pragmatism. Other, weaker and smaller, nations are not in that fortunate position. They are at the mercy of the whims of stronger countries as chillingly shown, for instance, by the ideas explored by Thomas P.M. Barnett (2004), *The Pentagon's New Map* or by the events described by Naomi Klein (2007) in *The Shock Doctrine*, or the strong-arm measures recounted by John Perkins (2004) in *Confessions of an Economic Hit Man*.

Concluding remarks: from hard to soft management

The aim of this chapter was in part to consider the necessity and feasibility of effecting a change in the overall framework that currently determines practices in economic and human development. The primary aim, however, was to use development as an example of a critical sector of human activity where the wrong diagnoses of the nature of the system resulted in costly and tragic mistakes.

An instance of such failure was glaringly exposed when world leaders came together in New York in September 2008 to discuss the Millennium Development

144 *Development*

Goals set for achievement by 2015 and adopted in the Millennium Declaration in 2000. The *Millennium Development Goals Report 2008* published by the United Nations sought to paint as bright a picture as possible but the sense of failure at the half-way stage could not be avoided. Tellingly, in his introduction to the report the Secretary General, Ban Ki-Moon, said:

> Some of the recent adverse developments reflect a failure to give these matters sufficient attention in the past. The imminent threat of increased hunger would have been lessened if recent decades had not been marked by a lack of investment in agricultural and rural development in developing countries. Climate change would be a less immediate threat if we had kept pace with commitments to sustainable development enunciated again and again over the years. And the current global financial turmoil reveals systemic weaknesses that we have known about – and left inadequately addressed – for sometime now.

The above quotation illustrates the other purpose for this chapter: to highlight the fundamental nature of the transformations that must accompany a move from orderly mechanistic formulations for development to ones based on complexity. The shift in viewpoint focuses on the multi-dimensional nature of the field and the linkages that exist between the large number of interacting elements that make development what it is; an evolutionary process that has no beginning or end and one that requires radically different tools from those adopted, with abysmal failure, in the past. The comments made by the UN Secretary General and the remainder of the report would have been blindingly obvious right at the start of the project in 2000 if a more appropriate formulation had been adopted. Instead, and yet again, an august body of experts has lamented the lack of past success and attributed the disappointing outcomes to a variety of factors that in total could be expressed as failure to recognise the nature of development as an adaptive complex system.

Such a realisation would have immediately pointed to Soft Systems Management styles that would have gone beyond the simplistic traditional remedies of yet more loans and financial assistance. Oddly enough, the Secretary General in the short quotation mentioned above refers to sustainable development and climate change; matters that were covered by the report of the World Commission on the Environment and Development set up by the UN in 1983. That final report was presented formally to the 96th plenary meeting of the UN General Assembly in December 1987 and agreed enthusiastically.[3] The report linked environment and development issues intricately and described development as principally a bottom-up activity that required an approach on a wide front. In many ways the commission's tools were in all aspects but name, similar to those associated with complexity.

However, the commission's report and its main recommendations, which received plaudits from all quarters, did not make a material difference to the way development is pursued; as is evident for example by the UN Secretary General's remarks quoted earlier. This is an intriguing phenomenon that is observed in other fields as well. It seems professionals know what they should do but somehow implementation does not accord with intentions. Fundamentally, the process of change from orderly- to complexity-based formulations imposes difficult demands that cannot be evaded. Cherished assumptions of predictability and control have to be abandoned. This is far from easy in an age where resolute management based on highly-specified hierarchical structures is taken as given. 'Soft systems management' styles; necessary in handling complex, *wicked* systems, could easily be seen as convenient means to shirk responsibility and accountability.

At the moment calls for involving 'frontline staff', 'local actors' and 'stakeholders' are seen as politically correct statements that must be, and often are, interpreted with a large measure of latitude. A move to a complexity framework in any field puts active engagement by one and all at the very top of the agenda. But such a departure has far-reaching consequences that might not be acceptable to those involved on the management side. For instance, most of the staff at the World Bank are located in comfortable offices in Washington DC, Geneva, and other desirable spots. They are also firmly in charge of the process. A shift to soft systems management would upset that arrangement. Local actors will assume greater responsibility and authority and World Bank staff would have to take part almost exclusively at a localised level. Naturally, that is not a change that would be wholeheartedly welcomed by one and all. The same could be said about development and funding agencies in the industrialised countries who feel, naturally, that they must be in charge for a variety of reasons including real or imagined local incompetence and corruption.

Clearly, a lengthy process of persuasion has to accompany advocacy for a move to complexity formulation. Soft systems management does not sit comfortably with top-down regulation and command-and-control: Where would that leave highly paid 'decision-makers' at the top of the hierarchy? And where would this leave the cosy notion that politicians in the centre are responsible for every facet of people's lives, large and small? Nevertheless, this shift is not an optional extra: if economic and human development is to be achieved in large areas of the world, including most of Africa, then more productive tools will have to be found. And achieving good outcomes in development are not optional extras either. Complexity demonstrates that development itself is not an isolated issue. Lack of development has well-understood negative consequences to all nations, rich and poor. As we will see in the next two chapters, terrorism and conflict are just two examples of aspects that are affected by poverty and bleak economic, social and political prospects.

8 Planning dreams into nightmares
The Iraq adventure

> Our goal remains a free and democratic Iraq that can govern itself, sustain itself and defend itself and is an ally in the war on terror.
>
> (President George Bush Jr)[1]

With such bright dreams, President Bush and his neo-conservative administration set forth on one of the largest planning operations in post WWII history – the invasion and then complete economic, social and political restructuring of Iraq. In the beginning, the signs looked good. The invasion went well. Many Iraqis were glad to be rid of a brutal dictatorial regime. They even managed to hold elections! Full transformation was just around the corner. And then, everything seemed to go wrong – warring factions emerged, the economy collapsed, basic services disintegrated, social order broke down, a hitherto secular society drifted into religious polarisation and a messy civil war emerged in which all sides sniped at the Americans. How was it that highly intelligent people and institutional structures, led and backed by the might of the reigning world hegemonic power, wound up creating one of the most expensive policy fiascos in modern history?

To explore this question from a complexity perspective we will *neither* provide a detailed history of US policy failure in the Middle East in general and Iraq in particular nor explore the various detailed reasons why the US pursued its policy in Iraq. We will also not discuss the moral rights and wrongs of the 2003 invasion and subsequent disastrous actions (a growing mountain of articles and books are doing that most eloquently). We do not see this as an enquiry into alleged incompetence or an investigation of who is to blame.[2] An operation that will ultimately cost the US several trillion dollars, killed and disrupted the lives of millions of Iraqis, tarnished America's and Britain's images for the foreseeable future, and has killed more Americans than the 9/11 atrocity is far too cataclysmic to be reviewed from the trivia of who did and said what. Instead, it is urgently necessary to ask more fundamental questions to explain why the whole project was doomed from the start, why it was impossible for decision-makers to change course, and how such disastrous miscalculations could be avoided in future. Therefore, this chapter will argue that:

- Assumptions of orderly, mechanistic thinking were evident throughout the Iraq debacle.
- These assumptions were by lengthy custom and practice embedded into the design principles on which most, if not all, US foreign policies were based.
- The fundamental belief that determines these policies rests on a view that with force; military, economic, and political, and good intentions, it is possible to redesign whole nations in a matter of a few years to bring them into line with European and American norms that took centuries to evolve.
- Even when it became abundantly clear that the wheels had fallen off the Iraq project the decision-makers found it impossible to change style: traditional thinking is difficult to shake off.
- The informed voices that cautioned against actions founded on such simplistic preconceptions were often based on 'common sense'. They were essentially ignored because they did not seem to come from a recognisable framework. Complexity could step in to fill this 'scientific' gap.

As demonstrated in the previous chapter, treating complex systems (such as whole nations with millions of citizens) as orderly phenomena amenable to practices borrowed from the industrial assembly line, is destined to produce disappointing outcomes, and in the case of Iraq, veritable disasters. In many ways, war is complicated but it is reasonably orderly. It responds adequately well to hierarchical command-and-control strategies based on hard management styles. It is, with few exceptions, finite with clear beginnings and ends. Peace, by contrast, is a complex process involving a variety of interacting elements which respond better to 'soft management' styles involving reiterative cycles that embrace consideration and reconsideration by mostly local stakeholders of both problems and solutions over lengthy periods of time. The process is ongoing, intricate, and highly messy. This vision of the peace phase was at odds with the underlying mission to rapidly impose the global neo-liberal economic model on Iraq and then radiate it to the rest of the Middle East. Fundamentally, the project was destined to be a nightmare from the start because decision-makers failed to appreciate the key difference between an orderly process, such as war, and a complex process, such as peace. Long-term peace is unwinnable by force alone; a fact that should be common knowledge based on historic precedent irrespective of whether one is conversant with complexity and its arcane ways or otherwise.

The predictability of failure

'Can-do' is a hallowed tradition in the American psyche. John Wayne conquered all obstacles. On the strength of this belief, augmented by huge resources and considerable knowhow, men were sent to the moon and brought back safely: a marvellous achievement. On becoming the only superpower after the end of the Cold War, it was not unreasonable to believe that the US could invade a country that had a ramshackle army, topple a hated regime, and then in a matter of months or at most a year or two convert the Iraqis to the best American traditions of democracy and prosperity. There were doubts, but elaborate US and British layers of government, and it must be said business, readily fell into line. Early congressional votes on action in Iraq were virtually unanimous and public support for Bush and action in Iraq were very high. Admittedly, spin and in one or two cases outright lies were deployed to win hearts and minds but that was simply the icing on the cake.

As things began to unravel, hindsight set in. The fallout on the political fortunes of some leaders was difficult to miss. An early casualty was Spain's Prime Minster José María Aznar who lost an election in March 2004. Berlusconi in Italy went next. The British tolerated Blair up to June 2007. Bush became the least popular US president ever, while the newly elected Barak Obama was left with the huge job of repairing the US's global reputation. But these were the insignificant personal losses that could not possibly be put side by side with the horrendous losses endured by ordinary Iraqis, Americans and Britons.

Unsurprisingly, there was a stampede of commentators looking for an explanation: the stupidity of George W Bush, the timidity of Blair, the rashness of the neo-conservative hawks, lack of post-war planning, inadequate invading force, and so on.[3] When failure became crystal clear, doubters simply wanted to withdraw forces either out of Iraq or at least to huge American bases in that country (the latter strategy being implemented in summer 2009). Those in power repeated the old mantras that sounded disturbingly familiar to those of the Vietnam era: 'Stay the course', 'One more push [surge]' and 'We will only leave when the job is done'. The same assurances, now peppered with 'good news stories' from Iraq persisted well into 2008 when US casualties hit the 4000 mark. When the security situation did start to improve in 2008/09 US troop numbers were reduced and the troops were taken out of the cities (the UK forces left Basra around the same time). But, even the newly elected President Obama would not give a clear date for the end of the US presence in Iraq. In short, no one knew what to do next. In the meantime domestic issues in the USA, a rampant economic recession, housing market collapse and escalating oil prices, eclipsed public concerns about the war that was a major factor in precipitating these negative economic trends.

Amazingly, despite the millions of words written and said, the nature of the systems involved and styles of management needed to deal with them effectively were hardly covered. The hot debate rarely ventured into the obvious point of why the project was stillborn. The high probability of the failure of a scheme to rapidly replan a whole nation, mimicking changes that took centuries to emerge in the US and Europe, was substantially ignored. What makes the whole tragedy and crippling losses unbearable is the fact that the mess was so predictable. In fact, it was the only predictable element in the whole affair.

Doubts were ignored

Expectations of failure are not 20-20 hindsight. Richard Cheney no less, secretary of defence at the time, spoke eloquently about the pitfalls that would have been involved in a move to invade Iraq at the end of the first Gulf War. It is instructive to repeat some of his observations as they forecast accurately the problems encountered after 2003:

> I think that the proposition of going to Baghdad is also fallacious. I think if we were going to remove Saddam Hussein we would have had to go all the way to Baghdad, we would have to commit a lot of force ... And once we'd done that and we'd gotten rid of Saddam Hussein and his government, then we'd have had to put another government in its place.
>
> What kind of government? Should it be a Sunni government or Shi'i government or a Kurdish government or Ba'athist regime? Or maybe we want to bring in some of the Islamic fundamentalists? How long would we have had to stay in Baghdad to keep that government in place? What would happen to the government once U.S. forces withdrew? How many casualties should the United States accept in that effort to try to create clarity and stability in a situation that is inherently unstable?
>
> I think it is vitally important for a President to know when to use military force. I think it is also very important for him to know when not to commit U.S. military force. And it's my view ... that it would have been a mistake for us to get bogged down in the quagmire inside Iraq.[4]

It is a pity Cheney was deaf to his own words when planning for the Iraq venture began in earnest in 2001. It is also a pity that George W. Bush did not seem to have read his father's and Brent Scowcroft's book, *A World Transformed*, about the Gulf War and other crises from 1989 to 1991. He would have found in chapter 19 wise words about the myriad of inter-related problems that would have cropped up following a move to Baghdad once the initial military aim of evicting Saddam from Kuwait had been accomplished ('Why We Didn't Remove Saddam', *Time*, 2 March 1998). They, as in the case of Cheney, knew the vast difference between a war with specific targets and the aftermath of war with its multivariate elements and their interactions that send an avalanche of baffling and changing signals to decision-makers. Cheney, Bush Sr and Scowcroft of course did not have the words to describe a complex system but in essence the link to complexity is unmistakeable.

These concerns were mirrored in two US government documents produced in January 2003, a couple of months before the war, and gave an uncannily accurate forecast of actual conditions that prevailed in Iraq once military victory was achieved. One was titled 'Regional Consequences of Regime Change in Iraq' and the other 'Principal Challenges in Post-Saddam Iraq'.[5] Later, similar conclusions were reached in possibly the most authoritative report produced by a Senate Select Committee on the Iraq mission titled *Prewar Intelligence Assessments about Postwar Iraq* and released for publication by a vote of ten to five.[6] The lengthy report does not mention complexity or the peculiar way complex systems behave that makes the use of force less effective, and often counterproductive, for so-called nation building. But the parallels with complexity thinking are again striking.

Other people at the top of the political tree, in Britain this time, were equally wary of a war in Iraq. Extracts published in early July 2007 of *The Blair Years* written by Alastair Campbell, Blair's leading spin doctor, suggested that most members of the Cabinet had misgivings. Several European leaders shared the same concerns but they were dismissed as 'appeasers'.

In short, the ensuing anarchy was predicted by many private and official elites and institutions. The public at large were equally concerned. London saw one of its biggest demonstrations in 2002 against action in Iraq. Expressions of doubt arose from an experiential and/or common-sense perspective as mentioned above. However, this sense of unease and uncertainty was not enough to overcome the growing number of government provided scare stories such as Saddam Hussain's supposed weapons of mass destruction (able to hit the UK in 45 minutes), alleged attempts by Iraq to obtain uranium and hints at links to terrorists. The onward rush to military action was virtually unstoppable.

It is only in later years that one was able to see the malaise that crippled US and UK policy-making as it related to Iraq. Fundamentally, there were a few people at or close to the top of the US administration, representing a broad and at times contradictory cocktail of interests, united by one unshakable conviction: that world affairs respond to decisive leadership backed by vast military power. Others were swept along, willingly and some less so, on a wave of optimism that quickly evaporated once the military battle was over. Some waited a little before breaking ranks but others decided to decamp once the agonies of so-called reconstruction phase came to the surface. Their observations and criticisms were almost identical and could be summed up as: 'Things went badly wrong, someone else messed up, but it was not me'.[7]

Comments made by Michael Bell, chairman of the International Reconstruction Fund for Iraq up to March 2007, are of special interest here as they come closer than most other observations to address the key issues that led to failure. He declared in May 2007 that reconstruction efforts are not only doomed but that they focus on the wrong projects such as high cost and high profile infrastructure schemes 'rather than developing institutions and human resources'. In addition he lamented that 'it was too much too soon' and that it was impossible to reconstruct under turbulent conditions of insecurity that followed the invasion.[8] Time and circumstances, it seems, were not on the side of those pushing the reconstruction efforts.

Even high-level military personnel such as Lt. General Ricardo Sanchez put forward precisely this assessment in a sweeping attack on US efforts in Iraq arguing that, 'National leadership continues to believe that victory can be achieved by military power alone ... The best we can do with this flawed approach is to stave off defeat.'[9] He then went to express the view that the process should have been approached on a much wider front. He could have added of course that Iraq, not unlike other nations, is a complex system and it would therefore only respond to appropriate styles of handling, but then probably he had not heard of complex systems. On cue the US military issued the latest version of the *Army Field Manual on Operations* in late February 2008. In addition to the traditional 'core missions' of offense and defence, a new one was added: 'stability operations'.[10] This would institutionalise the policy of 'nation building' by use of force that has been shown over and over again to be closer to mission impossible. Such is the strength of the present orderly paradigm. However, the manual does make one concession to more complex thinking: A 'Theatre Military Advisory Assistance Group' will be set up to help commanders in the field to better understand the culture of the people they attempt to 'stabilise'.

Was it just state incompetence and/or lack of preparation?

A simple depiction of the Iraq venture came to the fore in the years after the initial euphoria of the successful invasion in 2003. The story goes something like this: this was a huge bungle by the usual suspects led by an almost illiterate US president and a supine British prime minister. The usual suspects in this case were seen as neo-conservatives intoxicated by the writings of Leo Strauss and encouraged and funded by oil companies, arms suppliers, religious zealots and some Israeli sympathisers. This scenario implied two key elements: widespread incompetence by most of those involved in the planning and execution of the Iraq project, and the ability of a few individuals to bypass all the checks and balances introduced by the Founding Fathers into the American style of government and thereby to highjack policy-making within the US, and of course in Britain.

The above take on events is difficult to accept. The US 'government' is a vast structure. It was described by Samuelson and Nordhaus (1995: 302) in *Economics* as 'the world's biggest enterprise'. The proposal that it suffers from endemic incompetence, admittedly attractive to some, is unsustainable. Government service, particularly in the State Department and the Pentagon, attracts the top brains from the best universities. It would be curious that incompetence should suddenly pervade all levels of the hierarchy; not just in the US but in Britain as well. Furthermore, 'government' in the US embraces a wider field of expertise from business, research institutes, universities and think tanks. These are powerful and competent 'arms length bodies' that contribute to the work of government, which makes the suggestion of widespread ineptitude even less tenable.

Another diverting claim concerns the lack of advance planning for the war and, more importantly, for the crucial period after the war. This scenario is even less convincing than the first. Military planning for a possible invasion of Iraq was carried out over several years. Preparations for the period after the war were equally impressive. The 'Future of Iraq Project' was launched by the State Department in October 2001; significantly one month after the terrorist attacks on the World Trade Centre. The report was unclassified and published in May 2003.

152 Planning dreams into nightmares

Seventeen working groups were set up; each with ten to twenty Iraqi experts, up to five international observers, and moderators from the State Department. A list of the groups is given below to underline the point that all facets of the tragedies that assailed Iraq after 2003 were studied in advance and in depth.

- Democratic principles and procedures
- Economy and infrastructure
- Defense policy and institutions
- Education
- Public health and humanitarian needs
- Civil society capacity building
- Transitional justice
- Water, agriculture and environment
- Preserving Iraq's cultural heritage
- Public finance
- Oil and energy
- Local government
- Anti-corruption measures
- Foreign and national security
- Free media
- Migration
- Public outreach

The idea that the US and its allies went into the war in March 2003 in a state of total ignorance and without any thought of managing the peace are clearly unsustainable. Admittedly, decision-makers were incompetent in not appreciating the complexity involved in conquering and then stabilising a nation of well over twenty million people and with a lengthy and convoluted history. But then, few as yet appreciate this cardinal point.

Massive costs and negligible benefits

It is critically important to sketch the extreme lengths to which those pursuing US policy aims in Iraq went in their efforts to achieve their ends. The purpose here is to underline the power of current wisdom, based on centuries of 'custom and practice'. That wisdom suggests that planning for an end state in any field will always yield the desired results: it is only a matter of good preparation and adequate resources. The war, and the following lengthy period of so-called reconstruction, turned out to be exceedingly costly to everyone. Nonetheless, a change in direction was virtually impossible not only because of the likely 'loss of face', and the massive rewards that success promised, but because it proved difficult for leaders to understand the multitude of factors that combined to massively increase the odds against success.

The Congressional Budget Office, a nonpartisan outfit, released a report in July 2007 that put the cost of the war at $10 billion a month. Up to that point, the report estimated the cost to be about $500 billion (Rupert Cornwell, *Independent Online*, 11 July 2007). Joseph Stiglitz (the Nobel-prize-winning economist) and Linda Bilmes estimated the eventual cost of the Iraq and Afghanistan conflicts to be about $3 trillion. Comparison with other wars, including Vietnam, Korea, and past world wars, are simply astonishing (Stiglitz and Blimes 2008).

The financial cost of the Iraq project is of course insignificant when compared with the human cost that it inflicted. *Iraq Index* is the most authoritative report on casualties, military and civilian on both sides of the conflict. It is published and updated monthly by the Brookings Institution.[11] The June 2009 edition reported that US forces suffered 4316 deaths. In addition, it estimated that 31,354 American troops were wounded in combat.[12] Civilian and military losses on the Iraqi side are on an entirely different scale, but the figures are less reliable. *The New England Journal of Medicine* (31 January 2008) published a study carried out by the Iraq Family Health Survey Study Group which estimated the number of violent civilian deaths in Iraq from March 2003 to June 2006 to be about 151,000. Other estimates put the figure at 600,000 (*Washington Post*, 8 January 2007). Confirming *Iraq Index* figures, the UN reported that 'some 3000 Iraqis are being killed every month' (Patrick Cockburn, *Independent* 1 February 2007).

However, the pain does not stop there. The war resulted in civilian migrations and dislocations of biblical proportions. The International Organisation of Migration estimated that 5.1 million had been uprooted from their homes. Oxfam published a briefing paper on 30 July 2007, *Rising to the Humanitarian Challenge in Iraq*, that made grim reading. The case of the minorities, including the Christians who, the paper said, comprise between eight and twelve per cent of the Iraqi population, was underlined; including the drift of young women into prostitution in neighbouring countries simply to stay alive.[13]

Lavinia Limón, president and CEO of the US Committee for Refugees and Immigrants, selected 'The Silent Surge' as the title for her piece in the 2007 survey. She wrote, 'Two million refugees have escaped and fifty thousand a month continue to flee … [This] has been underreported in the media, elicited minimal response from the US Congress, and virtually ignored by the Bush Administration, European and Middle Eastern capitals, and the international community.[14] Significantly, the US accepted only several hundred refugees.

Two related aspects must be highlighted in the context of the above Iraqi deaths and dislocations: those affected are mostly the middle classes and the highly educated. This again was singled out in the Oxfam report mentioned above. In other words, Iraq has lost key people who could have played a significant role in the revival and reconstruction of the country. Within the wider picture, academics and doctors merit special mention. They became the target for an intensive campaign that effectively wiped out the top echelon of the intelligentsia in Iraq.[15] Again, according to the March 2009 *Iraq Index*, Iraq went from having 34,000 doctors before the invasion to 16,000 as of December 2008.

In addition to the obvious death and destruction caused by the Iraq misadventure, the indirect costs of the debacle will clearly affect American interests for the foreseeable future. Even its two most basic aims, the creation of a US-friendly government in Iraq espousing neo-liberal economics and control over Iraq's considerable oil and gas resources, were put in jeopardy by the events that followed the invasion. More worryingly, although the Obama administration is working furiously to repair it, America's reputation in Iraq, much of the Arab Middle East and some Islamic countries is as bad as it could be. Meanwhile, although there were elaborate efforts in Britain not to link terrorist attempts in the summer of 2007 to what has been done in Iraq, the evidence pointing in that direction is compelling. This, of course, does not excuse such acts but there is a need to consider why such attempts are being made at this point in time against Britain, as opposed to Austria for instance.

Moreover, it is important not to underestimate the harm the war has caused to America's reputation even within 'friendly' nations. This effect goes beyond revulsion at human rights abuses and torture that accompany most wars but were heavily reported in the Iraq debacle. Reputation loss in this latest episode concerns recognition of the waning power of US hegemony that reigned almost unchallenged since the end of World War II. This is, possibly, the gravest harm caused by a project intended to enhance Western, essentially US, supremacy. From a complexity point of view, this manifests itself every day in millions, if not billions, of micro-decisions – from deciding whether or not to buy US products, watch US movies, vacation in the US and/or send one's children to a US university – that are being negatively affected by the Iraq war and its legacy.

The underlying questions

Clearly, the costs, in every sense, were much higher than expected; in fact prohibitively so. Meanwhile, the main objectives became unachievable. Therefore, given sincere state effort and extensive planning, why did the US and its allies win the initial war so easily yet lose the 'peace', and why, as time went on, were decision-makers unable to change course in the face of mounting evidence that the whole project was falling apart? From a complexity perspective, the key problem was an inability to recognise the fundamental difference between the system that is 'war': how to conquer an opposing army; and the system that is 'peace': how to manage a whole country after it has been conquered. The two systems are dissimilar and hence require entirely different styles of treatment for them to be managed effectively and efficiently.

It is possible to argue that war is basically a predictable and orderly process. It might be, and often is, very complicated. But it is not complex. It is reasonably predictable and it responds well to command-and-control and hierarchical styles of management. A problem is identified, an objective is set, alternative solutions are examined, and finally a course of action is adopted. The 'waterfall method' (seen in Chapter 3) applied so successfully on an industrial assembly line applies equally well to war. Given determination and sufficient resources there is a high probability of success. The same applied even in the highly complicated conditions of a world war. When the US, with virtually unlimited resources, entered World War II on the side of the allies, Churchill recorded his now famous concept of the certainties provided by the 'application of overwhelming force'. Success, as he rightly predicted, was just a matter of time. Evicting Saddam's forces from Kuwait in the Gulf War was a predictable outcome. The 'Powell Doctrine' of overwhelming force was adopted by Bush Sr in that conflict.

Despite Clemenceau's warning back in the early part of the twentieth century that 'war is too serious a matter to leave to military men' it was nonetheless sensible to put the Pentagon in charge of the war in 2003. Given precise objectives by the politicians, 'military men' can be relied on to accomplish the mission. However, in this instance the Pentagon was presided over by a group of people, including Rumsfeld and Wolfowitz, who had their own specific ideas and agendas on how wars should be fought. In short, even at an initial stage of planning for the war proper there was conflict at the top. Essentially, Rumsfeld did not agree with the 'Powell Doctrine'. This initial conflict of opinions was covered at length by Bob Woodward in *State of Denial*. But the problem did not stop there. Rumsfeld and his close associates at the Office of Secretary of Defense were in turn advised by yes-men and others, such as the Iraqi exile Ahmad Chalabi, who had their own agendas (Phillips 2005: 67–76).

Rumsfeld had a concept of war that relied heavily on high-tech sophisticated weaponry that called for limited contribution by foot soldiers. He discarded the Powell Doctrine that called for overwhelming force to reduce the risk of defeat to a minimum. Rumsfeld decided the war could be won by using a force of 150,000 at most. What about the period after the invasion? That was thought to be easy as he and his aids were assured by people such as Chalabi that the Iraqis would welcome the American-led forces and in any case Chalabi had many thousands of Iraqis under his command who would swing into action once the US forces entered Iraq. These promises did not materialise, and the coalition forces proved to be too limited to hold the peace.

Basically, one camp, comprised of Rumsfeld and Cheney (in charge of Office of Secretary of Defense and the Office of the Vice President respectively), saw life as a mechanistic system: predictable and amenable to planning, pressure, force and control from the top. The other camp (Powell, in charge of the State Department) had a wider viewpoint that saw life from a more subtle and complex perspective that involved numerous interacting elements which offered limited predictability and required a more relaxed attitude to management that embraced a broader circle of partners.

There is of course no evidence that the individuals concerned had any understanding of, or interest in, the differences between orderly and complex systems. This would have been far too theoretical for these resolute decision-makers. Undoubtedly, individuals within the vast organisations mentioned would have heard of these systems, and some might have been experts in this field. Nonetheless, seeing the two warring camps in this sharp relief helps to explain how a war that lasted a few days and helped to rid Iraq of a hated dictator ended in a quagmire that devastated a whole nation and plunged the mighty US into a situation where it is seen as weak, incompetent and uncaring.

Where did the complexity come from?

This is a fundamentally important question. A look at war helps as an initial step. Two armies are about to fight each other for whatever reason. The mission is simple; acquire territory for instance. The general strategy would be worked out in advance but tactics on the field of battle would be more flexible to suit circumstances. If one of the armies is considerably weaker than the other, the system is made even simpler, and more predictable. That was the case on the eve of battle in the 2003 war. There were fears of surprises but these were factored into the strategy. The end result was reasonably predictable.

The 'complications' in the above situation concerned the movement of thousands of people and their equipment and in providing for their other needs such as hospital facilities, food and water, fuel, and so on. These logistical matters are routine aspects that have been rehearsed over and over. Contingencies would have been rehearsed in war games and scenarios of possible eventualities.

The Pentagon is possibly the best organisation when it comes to mounting a war. It is, like all other armies, hierarchical and it is based on rigid command-and-control. Over the ages these features have been known to provide best results. They did in Iraq. The war lasted a few days and Saddam's 'mighty' army capitulated virtually without a fight.

The 'peace' phase presents an entirely different picture. First, there were a wide variety of 'US interests' ranging from a general desire for someone, in effect anyone, to be punished for the events of 11 September to the president himself who not only wanted to complete a task that he thought was left unfinished by his father, the vanquishing of Saddam. Then there were the neo-conservatives, a loose collection of people with differing agendas. Many were driven by Leo Strauss' zeal to reform and rejuvenate America and in the process to reaffirm its rightful place as a great civilising power in the world. Even here there was a web of other interests. Prominent amongst these were a number of influential people who considered, from whatever motivation, US and Israeli interests to be one and the same. Virginia's Republican senator, John Warner, reflecting that school of thought, put Israel at the top the US 'vital interests' in the Middle East at a TV interview he gave in July 2007.

There was yet another very powerful faction that was happy to be affiliated to the neo-conservatives. This comprised the invisible men and women who believed implicitly in the goodness, at least for themselves and their businesses, of the global neo-liberal economic model. This faction itself was not undifferentiated as it included people from several backgrounds that agreed on the general aim but held differing views on the details; oil, defence contractors, arms manufacturers, construction and consulting firms, and so on.

The above snapshot of interests might seem complicated enough, but two highly significant policy groups merit special mention: State Department officials and Pentagon top brass. And these were not monolithic units either. The top brass disagreed, often violently, with Rumsfeld and his cronies at the head of the Department of Defence. Moreover, it is highly unsafe to assume that George

W. Bush and his vice president, Cheney, saw eye to eye on all, or any, of the core issues before, during and after the 2003 war.

The mission was initially simple despite the variety of 'US' interested parties: attack Iraq, remove Saddam, and then reap the rewards. However, the war was the easy part. After the military victory, however, the nuances between the groups on the US side, not to forget the British side, came to the surface. This was bad enough but once the coalition forces entered Iraq new local, regional and international actors entered the field with gusto.

The list of these additional participants is too extensive to be discussed in detail here, but a brief description will suffice to demonstrate the complexity of the system that those who initiated the war sought to convert or subvert to their way of thinking. Hanna Batatu wrote the definitive textbook, of 1283 pages no less, on the social, political, economic, tribal and religious make-up of Iraq's society (Batatu 2004). A graduate of Harvard with fellowships at Harvard and Princeton and with a lengthy teaching spell at Georgetown University, it is unlikely that key decision-makers and their advisors in Washington were unfamiliar with his seminal work, which was originally published by Princeton University Press in 1978. Sadly, that seems to have been the case, otherwise Iraq would have been approached more cautiously, or not at all. What makes Batatu's book even more pertinent is his description of the turbulent nature of Iraqi history and the resulting revolutionary movements that have gripped the country for decades.

In addition to this political mix, the religious map of Iraq is complex as well. The most populous zone to the south is the home of the Shiites, while from north of Baghdad to Mosul is predominantly Sunni. The Sunnis belong to four main sects. The Shiites themselves are not only Arabs; many are of Iranian origin, and they belong to a number of sects within the Shi'it creed. North of Mosul is still Sunni but the zone belongs mainly to Kurds rather than Arabs. The zone near Greater Baghdad and to the east is a mixture of theses Islamic groups but to add complexity the area contains Kurds who are also Shiites. Disputes between Sunnis and Shiites are the outcomes of historic events that stretch back hundreds of years. On the other hand, the disputes between Arabs and Kurds are relatively more recent but they are significantly more acrimonious. Further complexity is added to the Kurdish element because their area also includes Muslim Turkmen who have their own scores to settle with the Kurds. Iraq is also home to Christians, Arabs and non-Arabs, that crop up under various names such as the Chaldeans, Armenians, and Assyrians, Nestorians, and Syriac-speakers known as Syrian Orthodox (Batatu 2004: 13–36; Hourani 1991: 8–9).

Anderson and Stanfield put the topic in a nutshell: 'In selecting Iraq as the test case [to bring democracy and free enterprise to the Middle East] the US has chosen perhaps the most difficult case of all … In the complex tapestry of Iraqi society, traditional geographical divisions are reinforced by sectarian and ethnic divides.' (Anderson and Stanfield 2004: Introduction). The Americans might be excused for their rashness and naivety. However, those on the British side who collaborated with them should have known better on the basis of their lengthy and turbulent experience in Iraq.

A federal Iraq composed of three state-lets was advocated by certain groups in Washington for one reason or the other, but certainly easier control of oil was a primary consideration (Anderson and Stanfield 2004). The prospect of a Kurdish state at Turkey's doorstep was not an attractive outcome and the matter had become serious enough for Turkey to move considerable military forces to the border with Iraq by Summer 2007 and then to enter Iraqi territory in February 2008. Agitated Turkey to the north, simmering Iran to the east, and powerful Israel from further afield was enough to provide ideal conditions for mayhem. However, Syria to the west had its own agenda which most definitely did not include an untroubled occupation of Iraq by American forces and an influential Israel in cahoots with the Kurds on Syria's eastern front. Moreover, Kuwait on Iraq's southern border was clearly antagonistic to Iraq in the aftermath of the 1991 invasion. In short, the regional context for the Iraq venture was extremely complex and, as a result, highly volatile and this should have been clear to anyone with scant knowledge of the area.

Three acts of orderly reasonableness, but complexity madness

The brief description given above of the local and regional complexities should have been enough to convince anyone of the extreme risks associated with a war on Iraq with the intention of regime change followed by radical political and economic transformation that would become a model for others in the Middle East. Nevertheless, to demonstrate the difference between orderly and complexity perspectives on the occupation we will quickly highlight just three further decisions, among many, taken by leading actors in the US administration that helped to ensure total failure of a project that was already near impossible.

First, Rumsfeld presumably with the agreement of the president and the top brass decided a small force of some 150,000 would be sufficient; compared with 400,000 for Desert Storm. The assumptions, based on a simplistic view of the situation, were not implausible. Contacts had already been made with some of Saddam's senior officers and they promised little resistance in return for certain guarantees. US advisors, in addition, thought the Iraqi army could be used in the reconstruction effort following the war. In any case the superiority of US forces over Iraq's military resources was substantial. Furthermore, some of the expatriate Iraqis working closely with Rumsfeld and his close associates boasted that they have large numbers of followers ready to swing into action once American forces had secured Baghdad. As mentioned before, one such person was Ahmed Chalabi who was reputedly paid several hundred thousand dollars a month for this purpose. Ricks (2006: 36) accurately summed up his contribution as 'Chalabi's distorting effect'. In the event, only the Kurds delivered on these promises and that was because they had militias in existence for several decades (Feith 2008).

Second, on 16 May 2003, immediately after being appointed the presidential envoy to Iraq, Paul Bremer issued Coalition Provisional Authority Order Number 1: 'De-Bathification of Iraqi Society'. Previously, membership of the

Ba'th Party was required for all senior professional people in Saddam's era. By kicking them out of all senior positions, Bremer would in one stroke eliminate any legacy of Saddam. Even at the time, the negative implications of such an order were obvious: Bremer was advised by the CIA station chief, 'You will put 50,000 people on the street, underground and mad at Americans.' (Woodward 2006: 194).[16] With one stroke of his pen, Bremer removed the top echelon of Iraqi society with catastrophic implications for Iraq's ability to operate as a complex system.

Third, the minuscule size of the coalition force in Iraq was a subject of concern at the time and since. Bob Woodward (2006: 190) gave one example: 'James Dobbins, the post-conflict expert and State Department official ... [estimated] that 500,000 troops were needed in postwar Iraq ...' In this context, the third act of apparent madness was possibly the last straw. Bremer issued Coalition Provisional Authority Order Number 2 on 23 May 2003: 'Dissolution of Entities'; in effect dissolving the whole of Iraq's army and internal security forces. In the process several hundred thousand trained fighters, mostly with their weapons, were kicked out creating a substantial pool of recruits for the 'insurgency'.

From an orderly perspective all of these tactics would appear to be reasonable. For example, using our four golden rules of order:

- *Causality*: if it takes only 150,000 to win a war and gain control of Iraq, one can assume that it would only require 150,000 or less to maintain control. *Implication*: despite occasional doubters no more troops are needed.
- *Reductionism*: the enemy can be identified, separated and isolated from the general population. *Implication:* The Ba'th Party and army supported Saddam, therefore they are the enemy and can and must be eliminated from the political structures
- *Predictability*: once the Ba'th Party and army are dissolved and disbanded, significant resistance will cease. *Implication:* more troops are unnecessary
- *Determinism*: with help from Iraqi supporters (Chalabi, etc.) a new Iraq can be created with minimal US troop levels. *Implication*: trust Chalabi, and other expatriates brought in by the US, and more US troops would be unnecessary. America has right on its side and that makes the flow of future events practically inevitable.

Meanwhile, from a complexity perspective the absurdity of these presumptions becomes remarkably apparent:

- *Partial causality*: it may take only 150,000 to win a war in Iraq, but in the multifaceted aftermath of the war the picture changes radically. Managing a nation of over 27 million people in one of the most turbulent regions on earth requires completely different control arrangements. Leaving complexity aside, this fact has been thoroughly researched since the mid-twentieth century at least.[17]

160 *Planning dreams into nightmares*

- *Reductionism and holism*: some enemies can be identified, separated and isolated, but many cannot and some may change over time. *Implication:* large-scale de-Ba'thification (like large-scale de-Nazification after WWII) may be counterproductive, or at least a waste of time.
- *Predictability and uncertainty*: dissolving and disbanding the Ba'th Party and army will eliminate some forms of resistance in these institutional structures. However, other types of resistance/disorder (street protests, violence, etc.) may grow. *Implication*: do not assume that large scale de-Ba'thification will lower resistance.
- *Probabilistic*: it is probable that Chalabi, and others like him, had some support, but in the shifting sands of post invasion Iraq alliances, support is fluid. *Implication*: putting all your eggs in one basket is not just an old wife's tale.
- *Emergence*: evolving context promotes the growth and evolution of changing groups and interests. Both problems and solutions are shifting entities. Moving the situation from wartime to peacetime, creates new and constantly changing conditions. *Implication:* anticipating and controlling events in peacetime is extremely difficult to undertake using practices derived from wartime.
- *Interpretation*: as the impact of the aforementioned policy errors became apparent, Iraqis and other actors began to interpret the US administration in Iraq as inefficient and misguided. Initial support for the US began to fade. *Implication*: one should expect the costs of control to climb as the local population begins to feel that the US is no longer in control. Compelling a complex system, such as a nation of millions of people to behave in a given way, requires the application of overwhelming and costly force over very long periods of time; a near impossibility.

Could others have done better? The Stacey diagram

Einstein is reputed to have said, 'Only two things are infinite: the universe and human stupidity. And I am not so sure about the former.' The stupidity in the case of the Iraq war of 2003 was not the straightforward lack of brain power that Bush and others associated with the failed venture have been accused of by many authors. The standard criticism centres on their supposed inability to execute a project competently. There is an assumption that a more able group of people, presumably including the commentators, could have done a better job.

As discussed in this chapter, that completely misses the point. No one could have done better given the initial aim and orderly approach. The aim was not simply to rid Iraq of a hated brutal dictator by military force. The task was infinitely larger in scope: essentially to transform the social, political, cultural, and, above all else, the economic make-up of a country of some 27 million people with a rich ethnic and religious mix and a long and highly tortuous history. Moreover, the task had to be completed rapidly; the stakeholders demanded

Planning dreams into nightmares 161

quick returns. To accomplish the mission within that timescale there was only one way to proceed: adopt mechanistic styles of management planned, supervised, and implemented principally by military personnel using military tactics and tools. But Iraq, like any other society, is a complex adaptive system that evolves slowly and certainly not to order. Such a system can be nudged into a desired direction but at the end of the day patience and flexibility are needed and results cannot be guaranteed. Success relies on active local interactions mediated by simple enforceable rules that appeal to most of the population.

Essentially, therefore, the project was ill-conceived and doomed from the start. A different set of people at the helm would not, and could not, have done any better. Talk of the inadequacy of those who conceived and implemented the Iraq venture is the greatest deception in the whole episode. A way of visually recognising this is through a Stacey diagram presented in Figure 8.1.

As this figure clearly shows, the relatively easy part of the process was the war itself. Overwhelming force combined with clear goals and unanimity made it a perfect example of an orderly rational process that hierarchically organised techno-rational decision-makers are experts at carrying out. All Bush and the other elites had to do was put it in the expert hands of the military and the whole thing was over in days!

However, when we move away from this zone things immediately get more problematic. With the removal of anyone associated with the Ba'th Party and the dissolution of the army (as we saw in the above 'mistakes'), new ruling structures and elites in Iraq were quickly needed. Undoubtedly, there was a great deal of certainty and unanimity about the need for such elites and structures.

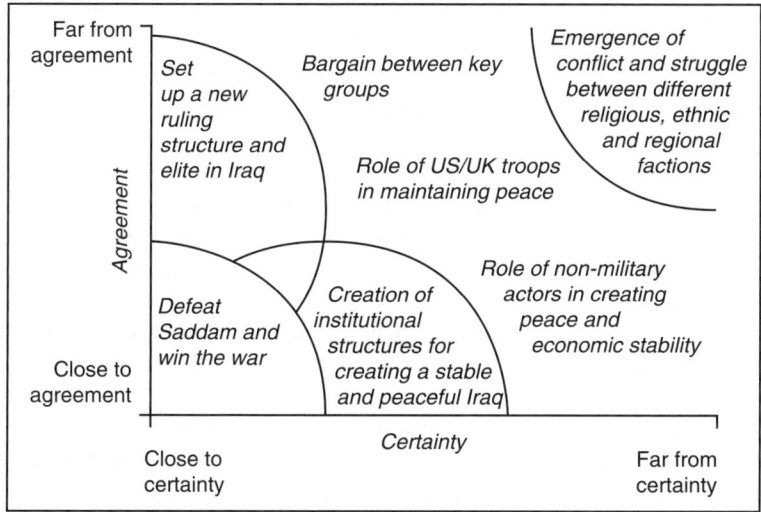

Figure 8.1 A Stacey diagram of the restructuring of Iraq.

Nevertheless, as one could easily imagine the degree of agreement over the structure of these institutions (democratic, semi-democratic, etc.), which groups would have privileged access to them (Sunnis, Shiites, Kurds, etc.) and the choice of particular actors would be very low since all of these are extremely 'political' decisions. No matter who was trying to bring these groups together, it would take time and rely heavily upon the relationships between the local actors. Bush (or anyone else) would only be able to do so much.

Similarly, even if there was general agreement over the desire to create a unified, stable and prosperous Iraq, what would be the best institutions to promote this? Should one choose a multi-party democracy based on a proportional electoral system that would represent all of the main social, economic and religious factions, but could lead to factional splits and divisions? Or, should one focus on a first-past-the-post (plurality) electoral system that would tend to favour broad coalitions, but could promote apathy within the general population? Should there be a privileged role for religious leaders? If so, which religions and which leaders? Obviously, this is a type of technical question where the best political scientists in the world would not be able to guarantee the outcomes of these choices. Once again, control by any external actor would be limited.

The picture gets even muddier when we move into the zone of fully complex systems where both certainty and agreement are mixed. Here obvious examples include: the evolving relationships between key groups or factions (Sunnis *versus* Shiites *versus* Kurds, etc.), the role of US/UK troops in maintaining peace and security (will their presence unite or divide the factions?) and the role of non-military actors (the plethora of economic interests, private security actors, international aid groups, etc.) in promoting peace and development. In this zone, the ability of the US/UK to greatly control the various processes and determine the overall outcomes was very limited, whether George Bush, Al Gore, Tony Blair, Paul Bremer was in charge or not.

Finally, as the conflict and civil war grew and emotions began to run increasingly high various forms of conflict and struggle emerged between the multiple groups and factions. In this situation overall security collapsed, normal social and economic relations broke down and those who could, fled (mass migration). In this zone, maintaining basic services became impossible (sporadic water or sewage supplies to Baghdad) and daily existence became a life-and-death struggle. Once the process had stepped over into this zone all actions and decisions of the US/UK elites became problematic.

Overall, from a complexity perspective, there are several lessons to be learned from the Iraq misadventure:

- *Know the system before you act*: pay attention to the type of system you are dealing with and alter your thinking and management styles appropriately.
- *The more complex the system and the goals, the lower the predictability*: recognise that beyond generalities, controlling and predicting complex systems is extremely difficult. Redesigning a country like Iraq, with all the obvious local, regional and international complexities involved, into a

Westernised neo-liberal democracy in a few short years was misconceived from the start.
- *Actors in the system are not interchangeable cogs and will themselves change over time*: you can't just briefly study and befriend a few actors in the system and assume you understand how the overall system would behave in future. Moreover, you can't just swap Ba'thists for non-Ba'thists and assume that the system will automatically function better and do so for the foreseeable future.
- *The limits of force and costs of ignoring interconnectedness*: force works in some situations, but is counterproductive in others. Encouraging a complex system to move into a desired direction of travel requires patience, time, and a full appreciation of the 'culture' of the system. Seeking to compel the system to instantly shift in an orderly direction is foolhardy at best. Likewise, interconnectedness plays a key role in shaping complex systems and determining the uses of force.
- *The importance of interpretation*: as is now common knowledge, winning the 'hearts and minds' of complex human actors is often more important than force in the long run. Their local interactions are paramount in determining how the system behaves and trying to encourage it to move in reasonable directions.

In the end, a complexity perspective may or may not have prevented the planning disaster of the Iraq project. Some argue that certain interests in the US and UK might have intended to destabilise Iraq on a long-term basis. However, knowledge of the way complex systems evolve and respond to external stimuli combined with some of the aforementioned lessons could have helped policy actors to avoid some of the awful consequences of the Iraq misadventure that have harmed not only Iraq but the regional and international political economy.

9   Exploding the myths of terrorism

> Terror is the greatest twenty-first century threat.
> (British Prime Minister Tony Blair, May 2003)
>
> The greatest threat this world faces is the danger of extremists and terrorists armed with weapons of mass destruction.
> (US President George W. Bush, September 2005)
>
> Terrorism is the greatest threat to world peace.
> (Russian President Vladimir Putin, September 2000)
>
> Terrorism is the greatest threat facing free democracies in the twenty-first century.
> (German Chancellor Angela Merkel, May 2006)
>
> No challenge is greater than the threat of terrorism.
> (Australian Prime Minister John Howard, May 2006)[1]

Given such unity among global elites on the importance and threat of terrorism, one would assume that terrorism was an easily defined and relatively simple phenomenon that merely requires the application of enough force to eliminate – a classic mechanistic system. But, is terrorism really that simple? If so, then the sensible response would also be simple: force in the shape of a 'war on terror'. The question is of fundamental importance. First, if terrorism is such a threat to world peace then it is vitally important to find effective counter-measures. Second, and equally important, the 'war on terror' as it has unfolded so far has brought in its wake a radical change in attitudes to individual liberty and human rights by governments of some Western countries presumed to be under threat from terrorists. Third, it is often said that terrorists' primary aim is to disrupt day-to-day life in the target countries. If that were the case then they seem to have succeeded. Life is now different from what it was only a few years ago. Irksome security checks at airports, arrest without charge for several weeks, identity cards, surveillance of emails, and relaxed attitudes to torture are only samples of the inevitable byproducts of the 'war on terror'. More fundamentally, the relentless focus on terrorism has increased the feeling of insecurity felt by citizens in countries considered, rightly or wrongly, to be potential targets.

In the aftermath of the awful events of September 11, a number of claims about terrorism were promoted in order to justify subsequent actions, including the 2003 war in Iraq. These claims rested on a bedrock of traditional orderly thinking that enabled key elites to suggest that they knew what they were doing and that they were in control. Basically, it was maintained that terrorism is widespread, that Western countries were on the terrorists' hit list, and that the only effective solution is to fight force with force. The Bush pre-emption doctrine was the natural corollary. First, denounce a country or a group as a terrorist instigator or supporter and then use force to eliminate the threat. This approach has involved the US and its allies, particularly but not exclusively, in several operations that have been not only been costly to the US in every respect but also counterproductive in terms of loss of reputation and the ability to influence events. This is a serious blow to its authority as the undisputed world leader and the defender of democratic norms. The price might have been acceptable, but the impact on terrorism has not been commensurate with the cost and moreover the publicity given to terrorist groups has given them unwarranted significance and glamour. Recruitment of operatives has been made easier.

In this chapter we seek to show that by ignoring all of the rules of complexity and wrapping themselves in confusion, and in some cases outright deceit, political elites using the inefficient orderly tools of the trade expended much effort and resources for modest or negligible results. Consequently, despite the efforts of many thousands of decision makers in vast bureaucracies, the war on terror was bound to fail.[2] Using a complexity perspective we hope to explore the real foundations of terrorism and explode some of the myths that have been propagated to justify and defend a whole range of interests and actions that have very little to do with the act of terrorism. Moreover, by viewing terrorism as a fundamentally orderly process and ignoring the law of unintended consequences the outcome of anti-terrorist actions, we argue, has been to create an increased sense of 'terror' and enhance the power of the 'terrorists'. We will illustrate these points by considering a series of myths that have cropped up as part of the sterile 'war on terror' project.[3]

Myth 1: There is a clear and agreed definition of terror, terrorism, and terrorist

As seen in the opening quotations, with so much agreement about the importance and threat of terrorism one would naturally expect a similarly high degree of agreement over the definition of terror, terrorism, and terrorist. However, even a quick glance at how these terms are used and abused shows that their definitions are marked by much more obfuscation and fluidity than clarity. For example, who is defining, when, where, and for what purpose determines the difference between 'terrorists' and 'freedom fighters' or 'nationalists' and 'insurgents'. We are already in a complex arena with many interacting protagonists and antagonists that is far from being orderly. This is not simply a matter of academic significance.

It has sufficiently serious implications that the UK government commissioned the highly respected Lord Carlile of Berriew Q.C. to independently review and report on *The Definition of Terrorism*, the title of his report, presented to Parliament by the Secretary of State for the Home Office in March 2007. Fundamentally, the report concluded that there 'is no single definition of terrorism that commands full international approval'. However, the report put forward some proposals to amend the Terrorism Act 2000 in an attempt to clarify several aspects while accepting that it is not practical to meet all objections; for example, actions in a 'just cause'.

The above proviso illustrates the difficulty in agreeing a suitable definition. At base, terrorism is a subjective term that is used to describe, pejoratively, actions by those of whom the user disapproves. In addition to activities by groups of people, there is, furthermore, 'state-sponsored terrorism' and 'state terror' and the same subjective quality applies to both. The *Oxford Concise Dictionary of Politics* gives a colourful definition that covers views about the actions of the Maquis against the Vichy government in France, the perception of George Washington as 'terrorist' or 'freedom fighter', and contrasting views by the Reagan administration of Libya's machinations abroad with US efforts in Nicaragua (McLean 1996: 492).

By contrast, politicians in countries that include the US, Britain, Russia and China, not to mention a whole raft of other smaller countries, strive to simplify the concept to suit their circumstances and interests. That difficult task of squaring the circle; turning a multidimensional phenomenon into a simple and orderly concept ends with importing more myth than reality into a feature of life that is both serious and costly to all concerned.

Myth 2: September 11 was a tragedy and not a political opportunity

In the days following the tragic events of 11 September 2001, Americans and much of the Western world craved a return to order. Seemingly unimaginable events, that many thought would have made a bad Hollywood script, were playing themselves out in 'real time' on television and radio sets across the globe. Who could have done it? Why did they do it? What would happen next? Desperate questions filled the minds of millions. The US government, despite some wobble in the first few days, quickly got into gear. Americans wanted answers and were willing their leaders to find them. A massive intelligence, police and military operation quickly swung into motion. Within a few days, the perpetrator was named: Osama bin Laden and the Al-Qaeda network. For some groups, terrorism filled an ideological vacuum left by the end of the Cold War and the collapse of the 'evil empire' of the USSR (Chomsky 2006).

The US public and government now turned to the question of what to do. For many, the answer was simple. Justice, Old Testament-style, must be brutal and swift. Bin Laden was in Afghanistan and the Taliban government would not give him up. Therefore, they must be 'taken out' so that Bin Laden and others could be made to pay. The Taliban's particularly nasty regime with multiple fundamental human rights abuses made the international public relations job relatively easy despite the fact that, according to some sources, they owed their rapid rise from obscurity in the early 1990s to Pakistan and Saudi Arabia. Some observers saw the Taliban as 'little more than proxies' for these close allies of the US (Burke 2004: 118). Support by the same countries, and the US directly, to Bin Laden during the Soviet occupation of Afghanistan does not require elaboration as it has been covered extensively by others (Heikal 1992; Ranelagh 1992).

A multinational coalition was quickly cobbled together. For a time it was fronted by Tony Blair desperate to make his name on the world stage, support the 'special relationship' with the US, and isolate his Conservative Party opposition. A 'war on terror' was duly announced and the various security forces began laying plans for activities at home and abroad. A bombing campaign combined with support for opposition Afghan groups soon led to the collapse of the Taliban. As of mid-2009, and despite a new US president, operations in Afghanistan continue without a resolution in sight. Over the years the campaign acquired a life of its own that seems divorced from the original impetus to start the war. As always, oil and geopolitical rivalries with Russia and China are not far from the surface.

Having seen the opportunities provided by the September 11 atrocity, the Bush neo-conservative-led administration in the US began to search for other 'terrorists' to, in the words of President Bush, 'smoke them out'. A lengthy list of 40–50 countries was bandied about with the usual suspects at the top: Iraq, Iran and Syria. By summer 2002, the focus was clearly on Iraq despite no evidence linking Saddam's repugnant regime to terrorist attacks on the US, or anyone else for that matter. There were other compelling reasons, as seen by those in power in Washington, to invade Iraq. In truth, terrorism and September 11 opened the door for several campaigns designed to achieve a variety of purposes that might or might not have had anything to do with the declared intention of protecting civilians against evil doers.

Within the US, the September 11 attack had a number of clear effects. It significantly galvanised the American sense of nationalism and patriotism. From the incredible outpouring of support for the victims and their families to the remarkable display of patriotic symbols (flags, colours, etc.) at virtually every political, social and sporting function, Americans were gripped by a nationalist fervour. This zeal had gone so far that even Karl Rove, President Bush's chief political strategist, noted that the reception of the president at Yankee Stadium during a 2001 World Series game was 'like being at a Nazi rally' (Woodward 2002). This intensely patriotic environment had clear political advantages for the Republican Party and President Bush. With a sluggish economy throughout 2002 and no major internal policy achievements, the aftermath of September 11 and the subsequent war on terror kept the media focused on Republican strengths. In the campaign for the November 2002 congressional elections, Democrats bitterly protested the Republicans wrapping themselves in a 'post 9/11 flag'. However, the tactic worked and the Republicans gained control of both Houses of Congress. In policy terms, defence expenditure expanded significantly, adding an extra $40 billion to the US defence budget, dwarfing the expenditure of any other national military system. Meanwhile, in late 2002 Congress passed the largest federal government reorganisation since WWII and created the Homeland Security Department with nearly 170,000 employees and an annual budget of $36 billion (*New York Times*, 20 November 2002).

The point of this brief review of the immediate events after September 11 is to point out that a whole range of political and policy actors, the President, Republican Party, military, US federal agencies, corporations, and so on, very quickly intensified their strategic and tactical activities turning a tragedy into a political and business opportunity.

Myth 3: One can reduce the elements of a complex international system into separate and manageable units

George Bush declared on 20 September 2001: 'Every nation in every region now has a decision to make. Either you are with us, or you are with the terrorists.' Reverting back to the Cold War years, the world was once again going to be made into a mechanistic orderly system composed of two distinct blocs. On one side were the forces of modern liberal democracy and on the other backward traditionalist fanatics jealous of Western and, in particular, US power and wealth. Those in each bloc are homogenous and their actions are both predictable and controllable.

Undeniably, most governments find it more convenient in explaining the terrorism phenomenon to blame creed and ethnic background or inbred hatred of democracy, than to consider the multifarious circumstances that turn ordinary human beings into ruthless killers. They also find it helpful to conduct the debate on terrorism in a fog of innuendos and half-truths. Essentially, all responsibility for 'terrorism' was and is placed firmly on the shoulders of the 'terrorists'. Policies and practices that might have been the cause of such action were deemed irrelevant. This principle is now enshrined in most anti-terrorism legislation. Seen from Slobodan Milosevic's point of view, this disingenuous philosophy has come a few years too late. Instead of facing a United Nations War Crimes Tribunal, he could have been the hero that battled against Islamic terror. Such are the fortunes of political life!

The reductionist approach, typical of a viewpoint based on an orderly perception of life, was found useful by a wide range of interests. Initially insignificant groups with grievances, real or imagined, and limited recruitment potential were suddenly able to recruit eager operatives from a much wider field based on religion or race. Bin Laden's campaign against the Saudi royal family and other regimes sponsored by the US became a more weighty movement fighting the cause of Islam no less.

Significantly, the basic structures of the international system hardly altered as a result of the new approach. Most countries acted in typically self-interested ways in dealing with the so-called terrorist threat. Pakistan quickly had its international debt renegotiated for its support of the Americans. Russia's support for the war on terror diverted attention from its war in Chechnya and gave Putin the opportunity to label the Chechen rebels as terrorists. China's support came just at the moment it was finally obtaining membership in the World Trade Organisation. Israel used it as a buttress for further repression of the Palestinians. Fundamentally, there is in practice no recognisable 'war on terror' that unites all nations who are supposed to be against terrorism. Equally, the whole of Islam, or whatever label you wish to put on the opposite bloc, is not intent on sponsoring or even condoning terrorism. Iraq was free from 'terrorism' and 'terrorists' until the so-called 'coalition of the willing' entered the country ostensibly as part of the war on terror. Obviously, separating the parts of the system into terrorist/anti-terrorists is an activity fraught with uncertainty.

Myth 4: The 'war on terror' is a global affair

This is basically a continuation of the preceding myth. There is simply no 'global' war on terror. Each country looks at events through the prism of its own interests and conditions. The 'war on terror' is an American innovation that emerged from challenges to American power, as perceived in Washington, and actions designed by a Bush administration marked by the presence of a number of individuals with distinct, and not necessarily similar, views on America's role in the world. The events of September 11 were the ideal moment for these diverse aspirations to coalesce into a simplified project that could be presented and justified to Americans and others. It was not surprising that Bush used the expression 'war on terror' in his first address to his fellow Americans. Within days, senior members of the administration and large sections of US media spoke of hardly anything else. The intention was perfectly clear: US security was the top priority and in addressing this aim the US administration would not be constrained by any legal or moral niceties. This is war after all! Dutifully, most countries supported this position in the beginning. In practice the support was observed only as far as it allowed each country to avoid US retaliatory action. For example, Pakistan's lukewarm participation in the so-called war on the Taliban, who were paid by Pakistan in 1994 to protect certain trade routes and regions, and against Bin Laden and in the most recent escalation of the fighting in Afghanistan in 2008/09, is a perfect illustration.

Despite frequent assurances of support, the war on terror is seen by many as an American affair driven by that country's wider global agenda. For instance, fears about a Pakistani atomic device falling into the hands of Islamic fundamentalists are only ever expressed in terms of the harm that could be done to an American city. The at least equally likel, scenario that the device might be used against India is totally absent from the West's reckoning. Again, the Obama administration and Brown government are trying to distance themselves from the slogan, but its liabilities remain.

Myth 5: Global and regional powers are simply reacting to terrorism

There are indications that the present preoccupation with terrorism is only the latest twist in a longer journey. America's position as the undisputed hegemonic power has been under threat since the late 1960s (Lake 1991). This reflects relative, rather than absolute, decline resulting from the emergence of competing regional alliances in Asia and Europe. As was the case when hegemonic power was transferred from Britain to the US, the process is lengthy, convoluted and messy. Scholars have argued, moreover, that a hegemonic power on its way down tends to adopt increasingly aggressive nationalist policies (Krasner 1982). John le Carré described this feature in his inimitable manner in an article in *The Times* of London in January 2003 when he wrote, 'America has entered one of its periods of historic madness, but this is the worst I can remember.'

The policies and actions of the Bush presidency were not prompted by madness of course. They reflected the behaviour of hegemonic power intent on keeping its place at the top of the global hierarchy for as long as possible. This is one plausible explanation for the increasingly belligerent foreign policies adopted by successive US governments in recent decades, including the so-called 'war on terror'. Other explanations, such as the wish to control Middle Eastern oil, can be seen as forming part of this broader philosophy.

The supposedly distinct and homogenous group of Islamic fundamentalists, it is repeatedly claimed, is driven by nothing more than irrationality verging on madness brought about by jealousy and religious antagonism. Actually, in most cases they are motivated by events and grievances that they have repeatedly articulated. These claims might not be acceptable or even justified but they are at some level rational. Bin Laden, for instance, has made his antagonism to the Saudi royal family perfectly clear. He has a quarrel with the US because it has steadfastly supported and protected, for perfectly understandable reasons, the Saudis. Arabs, Muslims and others, have made the focus of their hostility to the West equally clear: support for Israel and apparent or real lack of sympathy with the agonies of the displaced Palestinian people since 1948. Likewise, since the US/UK occupation of Iraq diverse terrorist groups emerged driven by ethnic and religious grievances against each other but united by the opposition to the occupation.

Each terrorist group has a degree of rationality behind it. It is not suggested here that all, or any, are justified in their complaints or the methods they use to draw attention to their cause. However, the pretence that all 'terrorists' are irrational and that the political, economic and social conditions in which they live have nothing to do with the actions and policies of the dominant powers cannot be sustained.

Myth 6: The 'war on terror' is being won

In the beginning, most commentators felt that the war was being won. However, as time has gone on many feel the war is already lost. In an address to the American Legion in August 2006, even President Bush began to show some hesitancy. An equally interesting matter would be to consider whether the 'terrorists', if one were able to treat them as a distinct group, feel they are winning. If we start with the obvious fact that terrorists are not fighting a symmetrical military war against the 'West' then the matter becomes easier to handle. It is safe to assume that at base they are trying to intimidate and disrupt nations in target countries.

The broader aim is ostensibly to be to force governments to modify their foreign policies.

According to the above yardstick the 'terrorists' have certainly won their first objective, ironically with much unintentional help from the Western governments concerned. People now do live in fear of terrorism despite the obvious fact that the risks to them are minimal at most. Their lives are also disrupted in a variety of ways. Did the governments change their foreign policies? The answer must be in the negative, so the terrorists have not won this element of their plan. But the price governments paid for this achievement has been hefty.

A lesson has been obvious though: aggressive policies run a distinct risk of making matters infinitely worse for all concerned, encouraging growth in conflict and terror. Specifically, such policies create a world in which US interests abroad, as well as ordinary Americans, would have to be promoted, and protected, by vast standing armies and an elaborate security apparatus overseas. It is virtually impossible to reconcile that setup with having amicable relations with other governments and nations. Terrorism has now become a cause and an effect: America cites terrorism as the reason for its aggressive actions abroad, and the terrorists describe their actions as the natural outcomes of America's foreign policies and aggression.

This paradox lies at the heart of the terrorism debate. On the one hand, we have US, and other Western, corporate interests that are nowadays closely intermeshed with political interests. This bloc wishes to keep the issue simple: 'Some people do not like us, because we are free, democratic, wealthy, or whatever, but we will go in and sort them out.' On the other hand, the facts, as argued here, reveal an entirely different scenario: 'Terrorism is a multi-dimensional complex phenomenon that has many causes and effects, and one that requires careful analysis and sensitive handling.' The focus on so-called Islamic fundamentalism, given prominence in Britain's National Security Strategy published in March 2008, was dissected in an article in *Foreign Policy* titled 'A World Without Islam' (January–February 2008). The author, Graham E. Fuller, who is a former vice chairman of the National Intelligence Council at the CIA, draws an uncompromising conclusion: a world without Islam 'does not present an entirely peaceful and comforting picture'. Why? Because there are numerous factors, ethnic, historic, religious ... that would lead to instability. Fundamentally, the picture is complex.

The problem lies in the contention that a 'war on terror' could be won or lost. The respected journal *Foreign Policy* in conjunction with The Center for American Progress polled over one hundred of America's top foreign policy experts; embracing conservatives, moderates and liberals (*Foreign Policy*, July–August 2006). It was reported that 'the results show striking consensus across political party lines. A bipartisan majority (84 percent) ... say the United States is not winning the war on terror.' This is not a surprising conclusion because the war could not be won. On the other hand terrorists cannot win such a war either. This was the underlying flaw in Bush's strategy as well as in Osama Bin Laden's. Fundamentally, there are many and diverse issues and grievances and they need to be tackled individually, fairly and constructively. Again, the recent changes in US/UK foreign policy seem to recognise this. However, these changes will take time and until a major shift is brought about the mayhem will undoubtedly continue.

Myth 7: Current terrorism is a new type of terrorism

A natural response to the above comments might be that we are entering a new era of increasing violence from terrorism and that the threat is escalating in absolute as well as comparative terms. This is incorrect: terrorism is an ancient phenomenon, and incidents and deaths associated with modern terrorism have been steadily decreasing since the mid-1990s. The declining trends will be discussed later, but a word or two at this point will shed light on the history of terrorism.

The average American, aided and abetted by the US government, would date the emergence of the terrorist threat as 11 September 2001. If pushed hard, he or she might go as far back as the bombing of the World Trade Centre in 1993. There are good reasons for this lack of historic perspective. First, until recently mainland America has been almost free from the impact of international terrorism, although US interests abroad have been a longstanding target. Second, the September 11 events were unique and overwhelming by any standard. Third, as discussed elsewhere, for diverse reasons the movers and shakers on the world stage found this foreshortened version of the history of terrorism convenient.

But what is the true picture? Despite the spectacular nature of the events of September 11, it only takes the briefest glance at the history of conflict and war to see that terrorism is an activity that has a multitude of ancient roots. Virtually every major war/conflict had elements of terrorism, or the use of terror against civilian targets, in it. Take your pick of conflicts or classical writers on international relations, Thucydides, Sun Tzu, Machiavelli, Clausewitz, and so on, and they will discuss how terrorist tactics were used by a variety of nation-states, political organisations and social groups.

More recently, few countries have escaped the spectre of mainly domestic terrorism in the post-WWII period. Europe and Latin America experienced a noticeable upsurge in the 1970s and 1980s; the Red Army Faction in West Germany, the Red Brigades in Italy, the Basque separatists in Spain, and terrorism in Northern Ireland are notable examples. Although there were a number of terrorist attacks in the US early in the twentieth century, the activity only became significant from the late-1960s onwards; the 'patriot movement' with which Timothy McVeigh (of Oklahoma bombing fame) is identified is an example. Diverse factors came together to cause this increase in the US, Latin America and Europe such as the Vietnam war, the Cold War, varied concerns about the environment and growth in government and corporate power, and of course a whole range of political, ethnic, and religious grievances.

Why is this myth significant? Because it turns what is essentially a limited and troubling feature that has existed for a long time in humankind's history into a crusade that sends ripples of instability through a vast complex system that has numerous actors of every conceivable shade of opinion and interest. This is highly risky when put alongside the technical advances available in the twenty-first century. The results are now clear to see on the world's turbulent stage. President Bush said in an address to Congress on 20 September 2001, 'Our war on terror begins with al-Qaida, but it does not end there. It will not end until every terrorist group of global reach has been found, stopped, and defeated.' This is a declaration of ongoing war that the US could not win by the orderly means of force. In fact it is an unsustainable statement that has imposed huge costs on the US and wider global community without the likelihood of tangible outcomes. The adventures in Iraq and Afghanistan require no elaboration apart from saying that US key interests; economic stability and growth, political dominance, security of stable energy supplies and safety of US citizens at home and abroad, have all been compromised by a strategy based on false, simplistic and orderly assumptions. Clearly, the Obama administration is responding to some of these lessons, but the escalating US involvement in Afghanistan indicates that their true importance has yet to sink in.

Myth 8: The scale of terrorism has radically increased

This is possibly the most glaring myth. Clearly, the threat from terrorism in all its formats warrants careful attention, but do the facts justify the official frenzy, and the antagonism that it inevitably generates, intentionally or otherwise, against certain ethnic or religious groups and nations?

Some of the most detailed facts on terrorism can be found in the 'Patterns of Global Terrorism' Reports that the US Secretary of State is obliged to annually submit. The reports go back to the early 1980s and are available on the US State Department's website (though since 2004 they have been replaced by individual country-level reports).

The 2001 report challenges the myth unambiguously, but unintentionally: 'Despite the horrific events of September 11, the number of international terrorist attacks in 2001 declined to 346, down from 426 the previous year.' This decline was not new and has been occurring since peak levels of the mid-1980s.

This trend was later confirmed by the 2003 report (Abbott *et al.* 2007: 41). In essence, after September 11 the normal pattern of virtually negligible deaths from terrorism resumed and has continued ever since. Meanwhile, the same reports pointed out that some countries in Asia and Africa suffer much higher levels of terrorist violence than the US and Western Europe have experienced in the post-WWII period. Interestingly, this aspect attracted relatively little comment among mainstream commentators.

Myth 9: Middle eastern and Islamic terrorism is a new and major threat

The State Department's 2001 report on international terrorism is uncomfortable about mentioning specific ethnic, religious and national groups. Having given a

174 *Exploding the myths of terrorism*

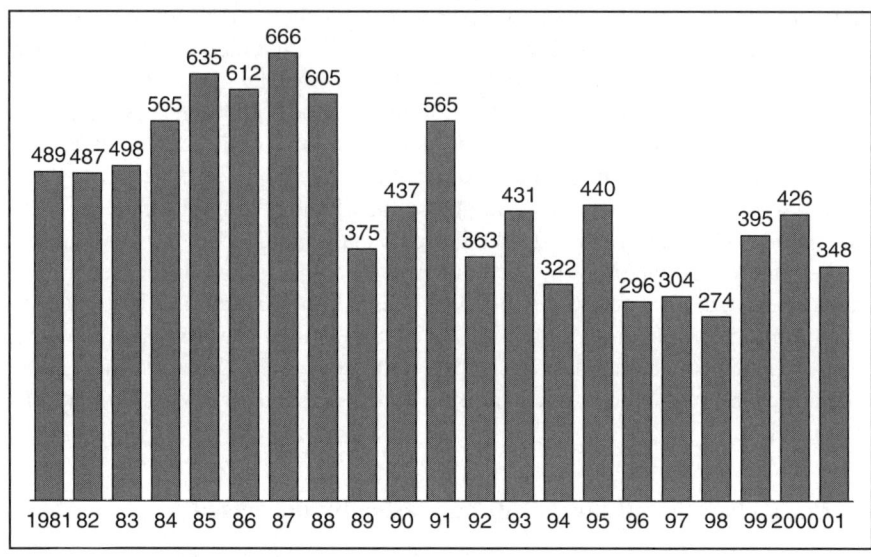

Figure 9.1 Total terrorist attacks, 1981–2001. (This chart is copied from Appendix 1, Statistical Review, Pages 171–176 of the *2001 Patterns of global Terrorism Report*).

disclaimer (p. xv), the report then goes on to suggest indirectly that Islamic groups (including the Palestinians) are the most significant offenders. This reflects the elite's sponsored view, which is gradually spilling into the popular consciousness, that Arabs and Muslims present the largest threat. But do the facts given in the report support that view? The Statistical Review again reveals a more balanced picture. Latin America comes at the top of the list by a long way when international attacks are grouped by region for all years within the period 1996 to 2001, as shown in Figure 9.3.

The same feature was highlighted in relation to specific anti-US attacks in 2001. The Middle East seems to be about average as a threat, especially when compared with the situation in Latin America. The 'Latin America Overview' section of the State Department's report is a document that should be made compulsory reading for media commentators. Put plainly, the US needs to clear its own backyard before it focuses its might on the Middle East and other Islamic countries. This is not said in defence of the terror activities of Islamic and Arabic groups, and there are too many of these. However, as numerous observers have already speculated, there is more to 'the coalition against terror' than straightforward concern about terrorism.

Middle Eastern terrorism is not a new phenomenon either. The Middle East has had its fair share during the last few millennia. Clashing religions, empires and peoples used various forms of terrorist tactics. The twentieth century was no exception. Zionist groups, such as the Stern Gang, Hagana and the Irgun,

Exploding the myths of terrorism 175

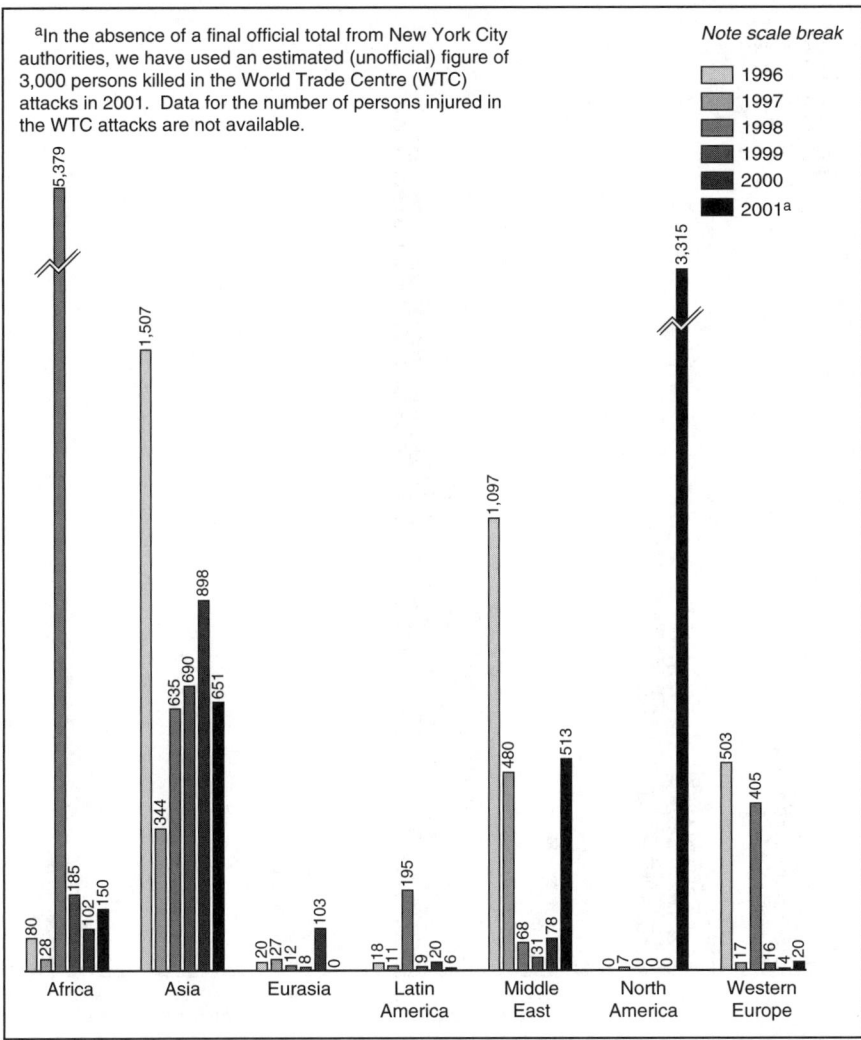

Figure 9.2 Total international casualties by region, 1996–2001. (This chart is copied from Appendix 1, Statistical Review, pages 171–176 of the *2001 Patterns of Global Terrorism Report*.).

struggling for the creation of Israel, and Palestinian groups, such as the PLO, Hamas and Fatah, fighting for the Palestinian cause have used a variety of terrorist tactics. Arabs and Jews are not the only participants in Middle Eastern terrorism. The Kurds also adopted terrorist methods for over a century against what they saw as their oppressors: Turkish, Iraqi, Syrian and Iranian governments (Zunes 2003: 20–21).

176 *Exploding the myths of terrorism*

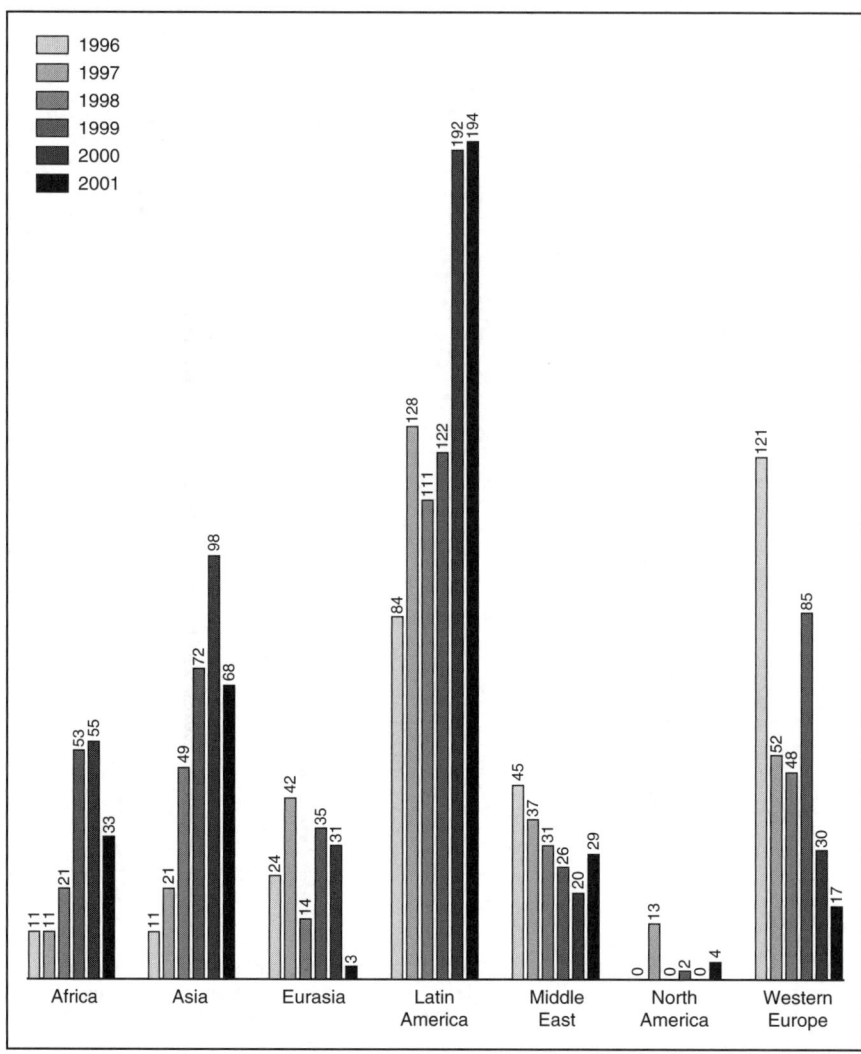

Figure 9.3 Total international attacks by region, 1996–2001. (This chart is copied from Appendix 1, Statistical Review, pages 171–176 of the *2001 Patterns of Global Terrorism Report*.)

Undeniably, many states indulge in terrorism as a form of 'diplomacy'. In 1954, relations between the West and Egypt were becoming too cordial for Israel's taste. On cue, several bombs exploded at a number of American and British establishments in Cairo and Alexandria. It was found eventually that the terrorists were not Egyptian nationalists but Israeli intelligence officers sent to Egypt for the purpose (Heikal 1992). More recently, in 1992 a military coup in Algeria nullified the first democratic elections in the history of Algeria since it gained its independence in 1962 when early polls suggested the Islamic Salvation Front would win decisively. The coup was strongly supported by the French government and was welcomed by the US government (*New York Times*, 24 July 1992). Not surprisingly, the aggrieved groups went underground and resorted to terror that has claimed thousands of lives in Algeria alone. Europe and the US have not escaped the anger of these groups either, as seen for example in the arrests of Algerians suspected of terrorist links in Britain in early 2003.

A key point: it is patently evident that terrorism is neither new nor limited to any particular creed, race or geographic location. When conditions are right any people of any religion could turn into 'vicious killers'. By the same token, no people are beyond redemption, as happened in Israel and South Africa, and as seen more recently in Britain in the acceptance of past IRA 'terrorists' as partners in government!

Myth 10: Al-Qaeda is new as well

Now this is something new, or is it? The US began to recruit and arm Mujahideen, to fight in Afghanistan more than six months before the Russians entered the country in 1979. Some of the funds were channelled through the government of General Zia ul-Haq who took over power in Pakistan in 1977. Over the next ten years the US, with some help from certain Gulf states and the so-called Safari Club, gave these Islamic fighters training, arms and facilities at a cost that exceeded three billion dollars. This includes the bases bombed by the US and its allies in 1998 and, more intensively, in 2001 (Burke 2004; Heikal 1992; Zunes 2003).

It was not enough for the US to recruit just any Muslims. They had to belong to the ultra-conservative sect known as the Wahabi tradition from Saudi Arabia. Most of the thousands of recruits came from that country, including Osama Bin Laden. He was a businessman who harboured broader political ambitions, mainly concerned with the need to topple the Saudi royal family. To that end, he developed a wide network of contacts in the early-1990s that became known in time as Al-Qaeda ('the base' in Arabic).

Clearly, Al-Qaeda is neither new and distinctive nor incomprehensible. Its evolution is a matter of record and America's role as midwife that aided its birth is unmistakeable. As always, the law of unintended consequences is in full evidence. Al-Qaeda managed to successfully stage a large-scale terrorist operation in the US, but its historical origins are not significantly different from other radical organisations in the Middle East. In fact for some, Al-Qaeda is a loose and disparate network of many groups that do not necessarily act in any recognisable format (Burke 2004). In short, it is not new and it is also not a monolithic group. Fundamentally, the so-called 'organisation' is part of a much larger complex phenomenon that requires appropriate handling if it were to be influenced to any tangible degree. Here as elsewhere, the wish to simplify 'terrorism' down to an easily transmitted concept to be deployed for a variety of purposes leads to adoption of orderly and therefore inappropriate lines of actions that generate unexpected and unwelcome outcomes.

Beyond the myths: what does complexity tell us?

Fundamentally, it tells us that many elites perceive national and international affairs as simple mechanistic systems that respond well to command-and-control management techniques. In essence, this is a natural consequence of the hierarchical setup inherent to the very presence of elites. The whole point of this management model is the ability of a limited group of people, presumably endowed with special acquired or inherited talents, to take decisions on behalf of the community at large. This, it has to be said, is often not unwelcome by the community. They are relieved to a large extent from the burden of making decisions on a whole range of often perplexing matters.

Lessons should have been learnt by now from the 'war on drugs' which is significantly more costly to the whole world, including the US. In this case decision-makers have tacitly agreed, after years of experimentation, that the problem is somewhat complex and it must be approached intelligently, with soft management techniques and considerable patience. The US's National Drug Control Strategy was expected to cost $14.1 billion in 2009 (see http://www.whitehousedrugpolicy.gov/publications/policy/09budget/index.html). Admittedly this only covered the Executive Branch but the costs were not unreasonable in the context of the dangers posed by drugs. The cost of the 'war on terror' is difficult to estimate accurately: figures vary from several hundred billions to trillions of dollars since its launch in 2001. The costs are mounting but the problem is not shrinking. People, especially in the West and specifically in the US, feel irrationally more threatened by terror than anything else, including drugs.

Undeniably, the threat from terrorism could not, and should not, be dismissed lightly, but it should be managed sensibly as a complex system. Furthermore, it should be seen in its true perspective as a serious but not critical issue of concern. Its use as a justification for covert or overt actions at home or abroad should be discontinued; especially as it is shown to be of such negligible utility. Businesses and governments often use risk management as a tool in determining policies. The likelihood of an event occurring is multiplied by the impact if such an event were to take place to determine the urgency or otherwise of taking action and the scale of the response. We would hazard a guess that application of risk management to terrorism would suggest a problem of relatively modest significance when compared with other risks facing society. For example, 'For women aged 19–44, domestic violence is the leading cause of morbidity, greater than cancer, war, and motor vehicle accidents.'[4] The government takes the matter seriously but the response is measured and well structured.[5] Similar statistics can be given for a multitude of life threatening activities. In the US for example, there are around 35,000 deaths per year due to guns and 35,000 due to automobiles. Terrorism in the US comes nowhere near these numbers!

If terrorism were shown to be a simple one-dimensional phenomenon with clear and linked causes and effects then the orderly, mechanistic solutions adopted by the US and others might stand a good chance of being effective. But the examples given above reveal a complex multidimensional activity that has numerous actors and interactions, and unpredictable inputs and outputs. Iraq was devoid of terrorist groups until the US and its allies stepped in and mounted a frontal attack in 2003 that was in part claimed to combat terrorism. It was repeatedly announced that the US is defeating Al-Qaeda in Iraq! This, and other myths mentioned earlier, demonstrates the self-defeating difficulties created when elites treat, intentionally or otherwise, complex phenomena as simple orderly problems.

A complexity map of international terrorism and its implications

Given the complexity of international terrorism it is relatively easy to construct a 'complexity map' of the phenomena, as is shown in Figure 9.4.

Beginning with the orderly aspects, as discussed above, international terrorism is not new and has taken a variety of shapes and forms over time. Nevertheless, the most predictable and orderly aspects of international terrorism is its root causes in international and global economic and political inequality and injustice. This unequal reality creates, on the one hand, the drive by the dominant state(s) and international actors to keep its hegemonic hold on world affairs, and on the other, a multitude of diverse grievances that affect significant groups of people that have no other means of redress but to resort to violence. The current phase of decline US hegemony is merely the latest form of this international tension. Unsurprisingly, when associated with poverty, despair and lack of justice, the mixture becomes explosive. Eminently justified, lamentations that no cause could possibly excuse

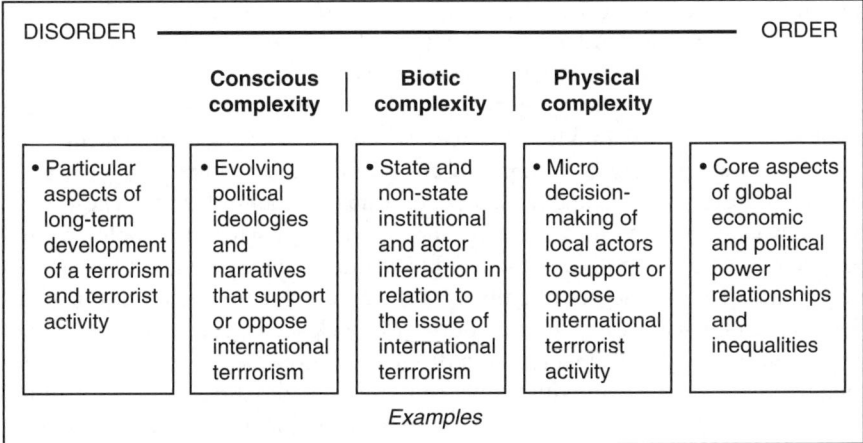

Figure 9.4 A complexity map of international terrorism.

the murder of innocent people are in practice a waste of time. The use of terror, usually as a last resort, to achieve cherished aspirations or to rectify grievances is not new and will not disappear simply because the dominant elites disapprove of the practice.

Moving towards aspects that mimic elements of physical complexity, one merely needs to think of the millions upon millions of daily micro-decisions that take place at the local level in the international system that support or oppose international terrorist activity. This can be a whole range of decisions moving from the obvious; does an individual decide to give a terrorist organisation money or not; do they vote for political parties that support the war on terror or not? At another level, these decisions may merely be small-scale economic decisions to buy or refuse to buy products from countries that support or oppose the 'war on terror'. The key aspect of this level is to see the 'flow' of these micro-decisions and how they generally stay within overall parameters. However, given a particular event or 'turning point', an avalanche of micro-decisions could lead to the toppling of governments or ruining of economies on either side of the conflict. The slow accumulation of opposition to President Musharraf in Pakistan that led to his resignation in 2008, despite strong support for the war on terror and help from the US, would be an obvious example.

At the biological complexity level one sees the interaction of states and non-state actors in an emergent process of adaption and survival. As discussed above, for many states the war on terror became a convenient excuse or pretext for repressing independence movements (Russia in Chechnya, China in Tibet, Israel in Gaza), demonstrating allegiance to the hegemonic power (Britain, 'standing shoulder to shoulder with the US), pushing for membership in an international organisation (Ukraine in NATO), obtaining new military bases (the US in several countries) or a whole range of other activities. All of these had interacting impacts on neighbouring states, internal political struggles and debates, and regional balances of power. Many of these outcomes were understandable, but also highly unpredictable, as daily events created new opportunities and threats to the multitude of actors. Added complexity could be seen by lowering one's gaze to the internal battles within states and societies.

At a conscious complexity level, international terrorism is clearly a struggle about the 'hearts and minds' of actors within the international system from the state elite who may be a target of an assassination attempt to a local actor scared to go to work on public transport because of bomb threats. These fears interweave themselves in a massive variety of ways with the ideologies and narratives of the terrorist, terrorised and bystanders. During certain phases there may be a remarkable degree of unity as in the case of the immediate aftermath of September 11 that led to a global outpouring of shock and disgust. However, within a very short period of time a plethora of interpretations began to emerge. These then interwove themselves with later events, bombing of Afghanistan, invasion of Iraq, creation of Camp X-Ray, and so on, thus taking on more subtle meanings.

Finally, at its most disorderly is the long-term outcome of terrorism and terrorist activity. Terrorist acts have triggered world wars (WWI), led to large-scale political repression (Indonesia), laid the foundation for new states (Israel), led to the breakup of existing states (former Yugoslavia), suppression of internal ethnic groups (China), external intervention by external actors (US in Iraq), and so on. Terrorists themselves have died in the act, been imprisoned, and occasionally integrated into a peace deal and become state leaders themselves. Clearly, predicting the long-term detailed impact and outcome of specific terrorist acts or the 'war on terror' is fraught with an extremely high degree of unpredictability.

Looked at from a complexity mapping perspective, the problem for the current period lies in the way that the US and its allies have decided to conceive of and carry out the 'war on terror'. Instead of viewing it as a complex system, they viewed it in a simplistic orderly fashion and applied traditional means to win the 'war'. The combination of its power and orderly vision blinded the US and it supporters to the problem of uncertainty. If you are uncertain, you don't have clear causation or blame. This implies that you can get Bin Laden and his network, but why attack states if they are only tenuously linked to the terrorist group. Also, how are you sure that your actions will 'stop terrorism' and/or put in a better regime? For example, why punish Afghanistan and Iraq and their peoples? The Taliban and Saddam Hussain certainly are/were bad, but how can you guarantee a better regime in the short or long run?

With hindsight, one can see that the US and its allies concentrated all of their efforts on an orderly attack and reordering of key terrorist states and actors (the Axis of Evil) using a coordinated military effort and lots of money. They easily won on the chosen battlefields, but immediately began to lose at all other levels. At a conscious complexity level, they quickly lost the battle for hearts and minds as the scandals of Camp X-ray, torture and mistreatment of prisons began to repel former supporters. At a biological level, former state allies began to waver in their support. Others began to demand limits on the war or developed a passive approach. Others took advantage of the situation to sort out a variety of other problems or issues. At the next level, millions of micro-decisions began to turn against the US as the 'war' went on. Voters began to throw out pro-US politicians. Tourists refused to travel to the US and international student enrolment in the US began to decline for the first time. These are obviously small costs in comparison to terrorist atrocities, but they accumulate over time and can become essential.

Finally, the one area that was left untouched was the one level where orderly interventions could have made a significant impact, the level of core global inequalities. To put it bluntly, the US and UK are currently spending many billions of dollars each year to fight the war on terror and its aftermath. The money will be spent. Soldiers will train and patrol new areas. New weapons systems will emerge. The Homeland Security Department will expand and the workers in it will work hard at their jobs. Much of the money will be re-circulated in the US economy and will act like a form of military Keynesianism to stimulate a sluggish American economy. Various threats will come and go, but all of this will only accidentally create more order and security unless it addresses some of the fundamental causes of terrorism: global poverty, inequality and injustice. From a complexity perspective, Americans and their allies are busily trying to reshape messy complex systems while ignoring the more fundamental regularities.

As a counter example, imagine if the US/UK took ten billion dollars a year, a fraction of what they are currently spending, and applied it to the fundamental causes, helping to eradicate starvation, improve poor water standards and provide basic health care in the poorer areas of the globe where most major conflicts emerge. In the Middle East, the key impetus for terrorism is the Palestinian/Israeli conflict. It is not beyond the wit of humankind to find a solution to this problem that has caused so much misery to both sides for over half a century. Imagine if this and other efforts were channelled through the UN and other international organisations to strengthen the role of the international community. Imagine if the US/UK supported an international court that could put real pressure on violators of fundamental human rights, instead of prosecuting only those who do not count as their friends. The exact outcomes would remain uncertain, but the general direction could be much better than the current situation.

To be fair, the US and UK governments have moved in these directions under the leadership of the Obama administration and we hope that this trend continues. Regrettably, there is little evidence to show that this change of style is caused by a better understanding of the complex nature of terrorism. Without such an understanding, it is only a matter of time before the next 'shock' event occurs and a new 'war on …' begins.

10 Conclusion

Towards a complex and humane public policy for the twenty-first century

Throughout this book we have tried to chart a new path in thinking about twenty-first-century politics, policy, and society. We sincerely hope that using a complexity framework may help social and policy actors to avoid some of the more 'orderly' horrors of the twentieth century. To conclude this book we would like to try and reply to some of the general questions audiences and our students raised over the past 10 years.

Isn't complexity just a recipe for doing nothing?

This question often comes from students or policy actors who see only the disorderly and unpredictable aspects of complexity. As we try to stress throughout, complexity is a combination of order and disorder, certainty and uncertainty, predictability and unpredictability. Hence, the search for boundaries, patterns and structures continues. Knowledge progresses, but the system does not stand still. It continues to evolve. In this sense, complexity is progressive (knowledge accumulates), but is also open (unknown developments always await us). In particular, for human systems what seems so fundamental and solid today often has a way of fading and rearranging itself tomorrow. To take just one example, if we were lecturing to undergraduate students in 1980 they would 'know' that the Cold War was going to be with us for the foreseeable future, nuclear weapons were the greatest threat to human existence, the European Union was a negligible political and economic organisation and China was an economic disaster. Today, the students 'know' that terrorism is going to be with us for the foreseeable future, global warming is the greatest threat to human existence, the European Union is the dominant European political–economic organisation and China is the next economic superpower. Who knows what they will 'know' in twenty years time.

Does this mean the end of progress? Are we back to Nietzscheian nihilism or Heideggerian fatalism in the face of forces beyond our control? This book is clearly focused on attacking the cult of order. We have chosen this focus due to its dominance in the twentieth century. However, complexity is an equal challenge

184 *Conclusion*

to the cult of disorder. That human beings cannot be gods, live in a symbiotic relationship with each other and nature and do not have complete control over their lives and hence complete freedom does not imply failure and apathy. As a leading complexity thinker, Klaus Mainzer, put it:

> The complex system approach cannot explain to us what life is. But it can show us how complex and sensitive life is. Thus it can help us to become aware of the value of our life
>
> (Mainzer 1997: 325).

Once one abandons the arrogance of order and despair of disorder and accepts the humbling limits of knowledge and uncertain potential which complexity implies then a new politics emerges: a politics of uncertainty, but also of openness, of mistakes and learning, of failure and adaptation. Exploring this new politics and the role of policy within it is what this book is/was all about.

Is there a morality of complexity?

From our understanding, though other authors explore this more thoroughly (Beinhocker 2007; Byrne 1998; Cilliers 1998; Mainzer 1997), there are core implications of complexity for morality, but openness is its major position. For example, if one were considering human rights, a complexity perspective would imply that the most basic human rights (rights to life, food, health, personal security, communication and expression) are universal and inalienable. These would come from the position that for humans to develop (like other complex systems), they must be allowed to survive (hence life, food, health and security) and interact with each other (hence individual and social freedom within a framework of basic social rules). These are the fundamental parameters of any human system – the orderly bit. However, if you move towards more detailed aspects of human rights (what kind of food, types of health, levels of security, degrees of freedom, etc.) the ability of a complexity framework to give clear guidelines quickly breaks down. For example, a complexity perspective will easily tell you that when the Taliban government in Afghanistan in the 1990s forced all women out of work, back into their homes and greatly restricted their interactions with other members of their society they were significantly undermining their own society's ability to adapt, adjust, evolve and develop. Likewise, in cases of civil war where the future is so uncertain that farmers are unwilling to plant their crops and hence threaten starvation for their societies, a complexity perspective will clearly see that the society's balance has been lost and is struggling to cope with too much disorder. However, as soon as one moves beyond these simple cases firm detailed positions are increasingly difficult to maintain. Is it 'good' for developing countries to put more effort and money into healthcare, education or security issues? In general, a complexity thinker would argue that this would increase the opportunities of that society to move in a more humane and sustainable direction. Hence, it would be morally 'good'. However, complexity stresses that one cannot know the exact nature of particular policies in any sphere. Hence, as detail increased and predictability declined, the ability of the complexity thinker to use a clear moral framework would wane.

What can complexity do for the weak?

We often get this question from our more idealistic students. At its root, there is nothing inherent in complexity to stop developed nations from exploiting the weak, powerful international/national economic actors from repressing the poor or for that matter technocratic policy actors imposing their orderly vision on a local community. In fact, the tendency of more complex systems to thrive mirrors the growth in international economic inequality that the world has seen from the beginning of the twentieth century onwards. This becomes very clear when one looks at the relative impact of global warming. Yes, rising sea levels from global warming will affect both wealthy and poor nations. However, wealthy nations like the UK, US, and so on, will find it relatively easy to adapt to the challenges, higher sea defences, different transport routes, new technologies to channel tidal wave power are all possibilities. For poorer countries, like Bangladesh, huge numbers of people will be driven inland and farming areas will collapse. The society will be put under huge pressure and possibly be made even more dependent than it already is. This is *not* to say that there is anything bad or inferior about the people in the poorer parts of the world. It merely recognises the obvious point that due to the less complex nature of their societies, there ability to adapt to the challenges of global warming is less than that of the advanced industrial countries.

Nevertheless, there is hope from a complexity perspective. It can do some things that may be remarkably powerful. First, it removes the veneer of scientific legitimation that is often used to justify particular 'orders'. Whether fascism, communism, capitalism or the next 'ism', there is no such thing as a final order! As such, it does not remove the power of orderly actors, but it chips away at their legitimacy and self-belief while encouraging less powerful and local actors to realise that they are the real forces in any successful system and their messy and complex reality is just as 'scientific' as the orders that are often being forced upon them. Linked to this distrust of final orders, complexity encourages diversity, interaction and expansion of learning and development in virtually every human social system. Giving local actors the ability to explore and adapt to their own systems is fundamentally more powerful than getting them to do as they are told from above. Finally, despite recognising how difficult it is to change their relative position, there is nothing in complexity that assumes that individuals at the bottom of the economic ladder are required to stay at the bottom. They are all full human beings doing their best to learn, adapt and adjust to their own distinctive complex situations, fundamentally no different from actors at the top of the ladder.

Why has complexity been so slow to spill over into the social sciences and public policy?

For many of our students and the policy actors who hear our presentations, complexity is so obvious and 'commonsensical' that they cannot understand why it has taken so long to reach these fields. In general, the social sciences have always had contesting traditions. As Thomas Kuhn pointed out long ago, the social sciences never stayed as tightly within their paradigms as the physical sciences did. Hence, as we pointed out earlier, there have been a whole range of challengers to the traditional orderly perspective. In essence, different types of complexity like thinking have been percolating within the social sciences for some time.

So, why focus on complexity and not some of these earlier trends? We have chosen to do this because complexity is one of the few radically interdisciplinary frameworks to intimately link the natural and social sciences. As such, we feel that complexity can have a much greater impact than these other approaches. Moreover, despite these multiple perspectives the dominant framework in the social sciences remains a traditional orderly one; just take a quick look at most modern universities where knowledge is rigidly divided between hierarchically organised disciplinary silos, also known as departments. Despite numerous calls for interdisciplinary learning and interaction, these structures continually inhibit this type of activity.

In the public policy domain the dominant framework of the twentieth century was undoubtedly the traditional orderly perspective. With this perspective, states could show that they were acting in the best, most rational interests of their citizens and that citizens should do as they were told. Moreover, citizens could even convince themselves that they were being taken care of by experts who knew what to do. Hence, the populace could wash its hands of individual responsibility in the relaxing allure of a belief in order. Complexity undermines both the dominance of elites and the passivity of local actors. From a complexity perspective, there are no guarantees of better answers, predictions and more certainty. Obviously, this is a very hard position to sell to a sceptical populace. Hence, complexity inspired approaches are often resisted by many political elites and distrusted by segments of the general population.

The good news is that many mid-range and local policy actors understand complexity immediately and intuitively. Struggling to balance the orderly demands of political elites against the daily complexity of local actors and their situations, local policy actors are often torn between the demands for order from above and by the requests for flexibility from below. Complexity appeals to them on two levels. First, as complexity thinking emphasizes, the main actors in any large complex social system are not the elites but the multitude of common actors making innumerable complex micro-decisions on a daily basis. From this perspective, local policy actors are more than just cogs in a large bureaucratic machine. In fact, their complex decision making often saves the larger policy systems and is more likely to create fundamental policy change than a few stern initiatives from above. Second, local actors are often caught between the 'order' of policy elites (demanding clear responses, outcomes and accountability) and the messy reality of dealing with normal complex people. Fundamentally, local policy actors are already experts in complexity. By embracing complexity thinking and its conceptual tools, they may have a better understanding of the nature of the overall policy system, improved tools for dealing with that system and the policy outcomes they are supposed to achieve and empower them to act in more flexible, responsive and adaptive ways.

Can complexity be inspirational?

We often get this question from those who begin to engage with complexity but have trouble being enthused by it. The quick answer is, in relation to its orderly predecessors it will always struggle to inspire. From a complexity perspective, there are no final orders, no happy endings and no ultimate resting points. Struggle, tension, difficulties and challenges are all a part of the process. They will never go away in human complex systems because they are a fundamental part of what we are. Learning, adapting, uncertain advances and unpredictable mistakes never end. Again, they are all a necessary part of the process and should be seen as such. Moreover, complexity does not even have a good rallying cry. Marxism had 'Workers of the world unite.' The best complexity can do is 'Be balanced!' And yet, for the foreign policy actor trying to find a way to manage the competing interests of rival groups struggling over scarce food resources or the healthcare administrator trying to coordinate limited resources between the needs of cancer patients, burn victims, children with diabetes and pregnant mothers, being balanced is the most reasonable tactic available. The key aspect that inspires us is how many local 'coalface' actors we meet who immediately grasp it and its implications.

How do we go forward from here?

This is the most tricky and practical question. At one talk that we were giving, a construction management expert raised his hand and said:

> I understand and agree with what you are saying. However, my clients want to know exactly how much a building will cost, what it will do and how long it will last. I know that all of these things are uncertain, but if I told them that and that all of my figures were probabilities, they would say that I didn't know my job and they would go to someone else.

At another talk a local health administrator, discussing her relationship to the central elites in the Department of Health, lamented:

> I know how complex and complicated my job is. I've lost count of how many central criteria I am supposed to hit. You try and tell them that it is too much and that they don't make any sense, but they refuse to listen.

188 *Conclusion*

> Or, even worse, they see you as a problem and you start to jeopardise your own career.

Both of these actors are caught in the same difficulty, being complex actors confronting elite and non-elite actors who want to believe that the system is fundamentally orderly. In the first case, the demands come from below, in the second from above. However, the pressure is the same. How can complexity help these actors go forward?

The truthful answer is that there is no magic wand to get out of this difficulty. In essence, they are at the cutting edge of a paradigm shift in our basic understanding of the social and natural world. Tensions between a complexity and traditional framework are an inevitable part of the process, as will be the tensions during the next paradigm shift. The key points to remember are that complexity is just as 'scientific' as the traditional orderly paradigm and that there are many others who are constantly confronting this tension. Exploring, responding to and managing this tension is an excellent learning experience. And, at a basic level, we are all experts in complexity. Our daily lives are full of complex, adaptive, emergent and interpretive aspects and decisions. We just need to realise that these are normal and healthy processes that make us what we are.

One should try to remember that rather than a Leninist elite or IMF/World Bank technocrats leading a radical and idealistic revolution towards 'the end of history', the cutting edge of the complexity transformation are the multitudinous daily micro-actions of local actors. It is these local actors, teachers, nurses, social workers, farmers, workers, etc., that are the real driving force for political and policy change. Given a stable framework and a reasonable degree of freedom, they will drive the system forward in a generally positive, but uncertain direction. They will, and should, make mistakes. However, it is this process of self-learning and self-development that will enable them to surmount the inevitable next challenges.[1]

And finally ... a couple of playful mental exercises

To end on a lighter note, we would like you to consider three playful questions that we ask our students:

- What would happen if we replaced the governors of the Bank of England with a room full of monkeys?
- What would happen if the UK tried to exactly copy the Norwegian primary education system?
- What would happen if a new super drug emerged that could suppress all of the symptoms of the common cold (or for our students a hangover)?

The quick answers from an orderly perspective are that: 1) the UK economy would collapse; 2) it should increase UK educational performance since Norway is commonly ranked significantly above the UK in primary education; 3) this would be a marvellous advance for personal health since the patient would feel better. However, from a complexity perspective things turn out to be a little different.

Regarding our Bank of England example, yes, if the monkeys were given total control over monetary policy the economy could fall into chaos. However, all complex systems have boundaries or a fundamental framework of rules that surrounds them. So let's put some rules around the monkeys regarding their control over monetary policy.

- First, we give them a switch and they are allowed to push it one way or the other once every three months (just like the current meetings of the Board). If they push it to the right, the UK interest rate will go up by half a per cent. If they push it to the left, it will go down by half a per cent. If they don't push it at all it will not change.
- Second, whichever way they push the lever the UK interest rate will stay within two percentage points of the average interest rate of the UK's major trading partners. Hence, the interest rate will never be significantly out of line with other developments in the global economy.

Now, what would one expect to see? Most likely there would be a greater degree of interest rate fluctuation but within boundaries. To respond to this:

- A cottage industry of 'monkey experts and advisors' who track the movements of the monkeys would quickly develop, analysing and providing advice on which monkey had a tendency to push the lever, at what time of the year, and so on.
- Individual actors would begin to build in strategies to deal with the added fluctuations. Banks would start providing 'monkey mortgage' protection for home owners (similar to fixed rate mortgages). Speculators would start betting on shifts in the monkey rate. Business might adapt by having 'monkey sales or specials' just before the monkeys were to set the new rate.

The key to this playful exercise is not to offend the governors of the Bank of England, but to realise just how good our systems are at adapting to complexity. All you have to do is take out the idea of the monkeys and replace them with 'normal' unpredictable international events, weather patterns and/or oil prices and you quickly realise that the governors aren't really 'in control' of the interest rate. So long as they act within reasonable boundaries, responding as best they can to the next inevitable shocks, we can all adapt relatively easily as well. In this sense, they aren't really in control of the interest rate and we (the average mortgage holder, car owner, businessperson, consumer, etc.) are the most fundamental actors in the system.

The second case, for years the English primary education system has been perceived to be in 'crisis'. For many, this is demonstrated by the relatively poor performance of English primary students in comparison to students in other advanced industrial countries in basic mathematics and language skills. From a traditional orderly perspective, there are four obvious tactics for solving this weakness: make the system work harder, make it work longer, input more energy into it and/or improve the system by copying bits from other, better performing systems. The first tactic implies eliminating unnecessary activities from the school day (physical education, art, music, etc.) and increasing demands for specific tasks and homework criteria. The second demands that one should increase the time spent in school. The third requires more money for the education system. Meanwhile, the last tactic implies all you would need to do is copy the aspects of the best performing system and your own system would immediately improve.

Under New Labour, the English education system experienced all of these tactics. The English now have one of the most structured and rigid teaching frameworks in Europe. English children begin school at the age of four (years before most other European children) and have one of the longest school years. Educational criteria and targets have also been developed by the Department of Education for nurseries so that the structured and targeted learning process can begin as early as six months after birth. Funding for education has increased substantially, and numerous attempts have been made to copy 'best practice' from other better performing systems.[2]

But quickly imagine all of the problems that would occur if one rigidly attempted to copy aspects of the successful Norwegian primary education system. First, the Department of Education would have to undo much of what it had just done. Norwegians don't start school until 6 or 7 years of age and have shorter school days. You would have to eliminate most of your national testing procedures (Norwegians have significantly fewer than the UK). Second, even if you tried to directly copy just one element of the system, say a Norwegian emphasis on team teaching at the primary level, this would have a whole variety of knock-on effects running through the English system (teacher working styles, administrative structures, children's attitudes to different teachers, etc.) that might undermine the overall outputs that you are hoping for. In no way does this imply that learning from other systems isn't possible. It merely emphasizes the reality of the distinctiveness of different systems and how learning from another system is very different from rigidly and mechanically interchanging bits of systems.

In our last example we ask what would happen if a new drug could be developed that would suppress all of the symptoms of the common cold (or for our students, all of the symptoms of a hangover). Viewed from a traditional orderly perspective, health is a simple state of not being ill. The sense of illness comes from feelings of pain, nausea, and so on. Therefore, eliminate the symptoms, the person would feel well and the problem is solved. Health is improved.

The obvious difficulty from a complexity perspective is that health is not a simple division between being healthy or ill. It is an interwoven process of emergent conditions and behaviours that we shape as we walk through our lives. In many cases, learning to adapt and interact with our bodies' signals (pain, nausea, exhaustion, etc.) are an essential aspect of finding a reasonable healthy balance in our own lives. Eliminating all the symptoms of a cold may make us feel better and enable us to work longer (as many adverts for painkillers claim). But in our complex reality, this may be the worst possible choice for our own health (when what we really need is to stay in bed and allow our bodies to recuperate).

In essence, the signals we are receiving are messages to rest, sleep, stay at home and recover. Repressing those signals may allow us to function uninterrupted in the short term, but it blocks our learning about our own health and may be very dangerous in the long term. Similarly, repressing the symptoms of a hangover may significantly lower the 'learning' that overindulgence with alcohol is dangerous and a major health hazard. Therefore, despite shrieks of dismay from some of our students, from a complexity perspective the pain and nausea of a hangover may be one of the most 'healthy' properties of consuming alcohol.

Notes

Introduction

1 Also note, the decline began in the early 1980s, long before the current AIDS epidemic which is pushing life expectancies even lower. Mortality rates of children under the age of five parallel these horrible developments. If the trends in mortality rates would have continued from the 1970s, about 40,000 children a year (around 10% of annual births) would die annually. Due to the changes from the 1980s onwards, over 85,000 (over 20%) die annually (UNICEF, *Trends in Child Mortality in the Developing World: 1960–1996*, UNICEF 1999).
2 Though his work was not purely orderly, reductionist and linear, Newton is often identified as the symbol of the Enlightenment scientific revolution. For a discussion of reductionism and contrasting approaches see Capra (1991), Coveney and Highfield (1995) and Hawking (1988).
3 Thomas Kuhn's thinking in *The Structure of Scientific Revolutions* (1962) was directed towards the physical sciences. He thought that the social sciences remained in a 'pre-paradigmatic' state and did not expect a normal science paradigm to emerge except in economics. Undoubtedly, the social sciences are messy and have a multitude of competing schools and theories that can entrench themselves in distinct educational institutions. However, we assert that there are dominant paradigms within the social sciences that structure basic theories and approaches and consequently public policy.
4 We use the term 'complexity' as a general term for a wide variety of non-linear theories which include various branches of chaos, complex systems and complexity theory. Traditionally, Ilya Prigogine is credited with sketching the possible applications of what he called 'dissipative structures' in the physical sciences as well as in society. He was awarded the Nobel Prize in 1977 for his contribution in this field. For a basic introduction to complexity theory see Cowan *et al.* (1994), Davies (1989), Kondepudi and Prigogine (1998) and Nicolis and Prigogine (1989).
5 Georgescu-Roegen was a leading advocate for moving the basis of economic thinking from Newtonian mechanics to thermodynamics. See Georgescu-Roegen (1999).

1 From orderly to complexity science

1 Epitaph intended for Sir Isaac Newton, *Oxford Dictionary of Quotations* (251: 26).
2 It is rumoured that he later regretted the now famous quotation.
3 For a philosophical discussion of the process of transformation, including the switch from linear to non-linear thinking, see Ferguson (1983). Hawking (1988: 1–14), on the other hand, provides an insightful technical analysis of the way scientific beliefs and methods changed through the ages. The Uncertainty Principle advanced by Heisenberg

had a particularly pivotal impact on the future course of scientific research. For a review of developments in physics see Davies (1987) and Peat (1991).
4 Major works on complexity include: Bar-Yam (1997), Capra (1991), Coveney and Highfield (1995), Gell-Mann (1994), Gleick (1987), Kauffman (1993 and 1995) Lewin (1999), Lorenz (1993) and Waldrop (1992).
5 The literature on the complexity paradigm and physical complex systems has now become quite large. Key works include: Nicolis and Prigogine (1989), Coveney and Highfield (1995) and Kauffman (1993, 1996). In addition, Waldrop (1994) and Lewin (1999) present an excellent general introduction to complexity.
6 Examples include: Bak (1997); Goodwin (1994); Gould (1989); Harris (2007); Lovelock (1979).
7 We are in full agreement with Scott except that we believe that democratic states can create equally horrific policy blunders when they are dealing with weaker external countries or groups (later chapters will make this point) or excluded internal groups or communities (e.g., American Indians in the USA, Aborigines in Australia)
8 For a discussion of the simple–complex dichotomy in Ancient Greek philosophy see Herrmann (1998).
9 This led Kant to conclude that 'The Newton (for) explaining a blade of grass cannot be found.'
10 For a review of the role of laws in the social sciences see Martin and McIntyre (1996).
11 It is important to note that post-modernism, by its own disorderly nature, has never been as structured and coherent as the modernist paradigm. Moreover, post-modernists have generally stayed at arm's length from the natural and physical sciences. Hence, the postmodernist critique has mainly occurred within the social sciences. Despite these limitations it has had a profound impact. For discussions of the development of the debate between these modernism and post-modernism see Bevir (1999), Bhaskar (1986), Byrne (1998) and Cilliers (1998), Delanty (1997) and Rasch and Wolfe (2000).
12 Cohen, Michael D., James G. March, Johan P. Olsen, 'A Garbage Can Model of Organizational Choice' *Administrative Science Quarterly*, Vol. 17, No. 1. (Mar., 1972), pp. 1–25.
13 For an excellent overview of the spread of complexity theory and a critical review of its popularisers see Thrift (1999).
14 There are a number of edited works that are starting to explore complexity and policy. These include a 2008 special issue on 'Complexity and Public Policy' in the journal *Public Management Review*, M. Wallace, M. Fertig and E. Schneller (eds) (2006) and Dennard, Richardson and Morcol (eds) 2008. The journal *Emergence* is also an excellent resource.

2 Concepts of complexity

1 For readers wishing to read more about terms and concepts used in complexity science we recommend the appendix to Coveny and Highfield's *Frontiers of Complexity* (1995). See also Cilliers (1998) and Casti (1994).
2 For a discussion on turbulent flows see Gleick (1987) and Briggs and Peat (1990). For compression see Coveney and Highfield (1995).
3 A full discussion of the subject of attractors is beyond the scope of this, essentially practical, book. Their arcane shapes and characteristics, not to mention the dimensions in which they exist, can be found nowadays in many textbooks as well as the Internet. See, for example, Casti (1994: 29–30), Gleick (1987: 137–44) and Mainzer (1997: 273–75).
4 For an excellent description and demonstration of the butterfly effect and attractor see: http://www.exploratorium.edu/complexity/java/lorenz.html
5 Malcolm Gladwell wrote a most informative and readable book; *The Tipping Point: How Little Things Can Make a Big Difference* (2001), which became a best seller.

He selected events from recent history to illustrate the propensity for human activity systems to change attractors when least expected.
6 It is perhaps helpful to highlight another point in relation to this example: the communist ideology did not have sufficient variety to withstand changes in the global political economic system. It proved to be less stable than the liberal economic ideology that managed to adapt and reform in line with the changing environment.
7 We will not explore or use non-linear mathematics or fractals in this book. However, it is essential for the reader to realise that both form the basis of studying and modelling a huge variety of physical, biotic and conscious complex phenomena including the stock market (Robert Savit, 'Chaos on the trading floor' in Hall, N. (ed.) *Exploring Chaos: A Guide to the New Science of Disorder*, Norton: New York, 1991). And in the programming for computer games such as SimCity, see Johnson (2002).
8 For a general discussion on non-linear mathematics, see Mainzer (1997), Prigogine and Stengers (1984), and Stewart (1989).
9 See Goodwin (1994), Harris (2007), Taylor (2005).
10 Mendel (1822–84) did his best to overcome that drawback by studying fast growing garden peas in 1865 to develop his laws of heredity.
11 Ray defined local rules for interaction and added a killer 'reaper' to enable fitter organisms to monopolise the available resources. The entities soon mutated into a diversity of new creatures complete with parasites. A form of community life eventually appeared (Coveney and Highfield 1995: 253).
12 One evolutionary theorist, William Hamilton, speculates that sexual reproduction emerged as a strategy for organisms to defend themselves against harmful parasites that found it harder to adapt to the diverse attributes of a population generated by sexual reproduction (see Gell-Mann 1994: 253).
13 The classic text on exposing the tendency of reductionist and determinist biology to distort the complex nature of humanity is Gould (1981).
14 Similar concepts have been bundled into a general concept of emergence, for example see Johnson (2002), Sawyer (2005) and Stewart (1989). We have chosen to use them separately in this book and group them under the framework of complexity.
15 Eldredge of the American Museum of Natural History and Gould of Harvard University put the concept of punctuated equilibrium forward in 1972. Although the idea is generally accepted it has had its critics as well, for example see Coveney and Highfield (1990) and Gell-Mann (1994: 238).
16 At a fundamental level, as the arrow of time proceeds, new regularities are added to the system and the statement needed to specify it becomes progressively longer indicating an inherent tendency for complexity to increase (Kauffman 1993: 232). As stated pithily by Gell-Mann: 'The evolution of Earth, of the weather on its surface, of the prebiotic chemical reactions that led to the emergence of life, and of life itself, all illustrate the accumulation of frozen accidents that have become regularities for restricted regions of space and time. Biological evolution, especially, has given rise to the emergence of higher and higher effective complexity' (Gell-Mann 1994: 231).
17 This is obviously a huge field ranging from cognitive psychology to traditional debates over the mind–body duality. Chapter 4 in Mainzer (1997) provides an excellent introduction to the history of some of these debates and the position of complexity science in relation to them.
18 References on the topic of elites and hierarchies are plentiful. See, for instance, Marger (1981), Bottomore (1993) and Lerner et al. (1996).
19 For a full discussion of emergence see: Johnson (2002).
20 For a technical discussion, including predictability in political modelling, see Kravtsov and Kadtke (1996).
21 There is a growing literature on complexity and organisational studies. The founding father of this literature is Professor Ralph Stacey.

3 Tools of complexity

1 There is also a growing range of modelling tools (multi-agent systems, network and neural modelling) that have emerged to explore complex systems. In order to keep this book at a less technical level we will not be exploring these areas. However, we would recommend the section on 'Multi-agent systems and complexity' edited by Peter McBurney in Bogg and Geyer (2007) as a brief introduction to some of the work in the field and the work of Robert Axelrod on the application of this type of modelling to political and policy issues (Axelrod 1997; Axelrod and Cohen 2000).
2 For a discussion on the relationship between complexity theory and artificial intelligence see: Coveney and Highfield (1995), Holland and Langton (1992) and Mainzer (1997).
3 Equating up with fitness and down with lack of fitness is an arbitrary designation. It could easily be reversed as Gell-Mann (1994) does.
4 'Dilemmas in a General Theory of Planning', *Policy Science*, Vol. 4, No. 2, June 1973 as quoted in *Tackling Wicked Problems: A Public Policy Perspective*, Australian Government Public Service Commission, 2007.
5 'A Voice for All: Engaging Canadians for Change', Institute of Governance, www.iog.ca.
6 Peter Checkland, professor of systems (emeritus) at Lancaster University, is credited with developing SSM in the 1960s although others, including Brian Wilson, have also made major contributions to this field.

4 Politics

1 The rise of this rationalist/orderly position is charted brilliantly by Green and Shapiro (1996).
2 See Gell-Mann (1994), for a discussion of 'depth' in complex systems.
3 For further reading on the topic of elites and hierarchies see Marger (1986), Bottomore (1993) and Cassis (1994). For a critical look at elites in the US see Lasch (1995). Lerner *et al.* (1996), presents an interesting concept of 'strategic elites' as an explanation of the pattern in the US.
4 See: 'Just enough to keep aid-givers happy.' *Economist*, 23 November 1996.
5 For some, particularly authors from a corporatist tradition, this poorly organised European level was a significant threat. Corporatism was not feasible at the EU level, because 'the political system of the post-1992 European Community will be fundamentally and by a quantum leap *more complex* than anything that has preceded it' (Streeck and Schmitter 1991: 151).

5 Health

1 See www.globalcomplexity.org/Costversusvalue.htm
2 See World Bank documents in www.worldbank.org
3 See www.hsph.harvard.edu/review/review_summer_02/txt677cuba.html
4 For a more detailed discussion of the complexity of this case see Rihani (2010).
5 The British health system is divided into four sub-national health services: the Scottish, Welsh, Northern Irish and English Health Services.
6 John Mohan, http://www.historyandpolicy.org/papers/policy-paper-14.html
7 Note from E.R. Griffiths to Secretary of State for Social Services, http://www.sochealth.co.uk/history/griffiths.htm, accessed on 18 February 2009
8 See http://www.bristol-inquiry.org.uk/final_report/report/sec1_chap_4_7.htm
9 http://www.nhs.uk/Tools/Pages/NHSTimeline.aspx
10 For further details on ISTC see Player and Leys (2008).
11 http://www.dh.gov.uk/en/Publicationsandstatistics/Publications/PublicationsPolicyAndGuidance/DH_4086665

12 http://www.dh.gov.uk/en/Publicationsandstatistics/Publications/PublicationsPolicyAndGuidance/DH_085825
13 *Transforming Community Services: Enabling New Patterns of Provision*, Department of Health, Gateway Reference 10850.
14 See Bogg and Geyer (2007), Cooper and Geyer (2007), Cooper and Geyer (2008), Cooper *et al.* (2004), Fraser and Greenhalgh (2001), Gatrell (2005), Holt (2004), Plesk and Wilson (2001), Simmons (2003), Steinberg (2005), Sweeney and Griffiths (2002) and Sweeney (2006).
15 *Working in Systems*, published by the NHS Modernisation Agency in April 2005.
16 See for instance 'Ten years of going round in circles', *The Sunday Telegraph*, 25 February 2007, and 'NHS reforms-a case of déjà-vu?', http://news.bbc.co.uk/1/hi/health/4603740.stm
17 Nic Flemming, *The Telegraph*, 12 October 2004.
18 The *British Medical Journal*'s 'Editor's choice' in the 7 June 2008 issue was 'Complexity theory' and this referred to other recent papers on the same topic. See http://www.bmj.com/cgi/content/full/336/7656/0 See also A. Shiell *et al.* 'Complex interventions or complex systems? Implications for health economic evaluation', http://www.bmj.com/cgi/content/full/336/7656/1281. Moreover, The King's Fund, Nuffield Institute and NHS Leadership Centre held a conference in January 2003, 'Forward Thinking' on complexity in healthcare and a couple of months later, in March 2003, the Mayo Clinic in the US organised an event on 'Complexity Science in Practice'.
19 See for instance, J. Glasby, E. Peck, C. Ham and H. Dickinson, 'Things can only get better?-the argument for NHS independence', Health Services Management Centre, School of Public Policy, University of Birmingham, http://www.newscentre.bham.ac.uk/documents/Things_can_only_get_better–the_argument_for_NHS_independence.pdf
20 http://www.guardian.co.uk/politics/2006/oct/09/conservatives.uk1
21 NICE Health Technology Appraisal on Patient Education, http://www.nice.org.uk/page.aspx?o=68326. [Accessed Oct. 2006]
22 Further applications of tools to complexity to diabetes management are explored in Cooper and Geyer (2007, 2008, 2009).

6 The international arena

1 Lynn and Jay (1989: 171)
2 Leon Lindberg, one of the early researchers in European integration captures this trend perfectly. After writing several major works on European integration in the 1960s he turned more towards studies of inflation and comparative political economy in the 1970s and early 1980s, returning to EU studies in the late 1980s.
3 See Barry and Walters (2003), Cederman (1997), Dunn (2007), Harrison (2006), Grande and Pauly (2005), Jervis (1998) and Richards (2000).
4 Comparative politics has generally had a tendency to be less orderly than international relations. Comparative politics never had a unified core theory and has always found it problematic to apply reductionist frameworks (Hague and Harrop 2004).
5 See European Commission (2001) *Brussels: Capital of Europe – Final Report*. Available at http://www.pedz.uni-mannheim.de/daten/edz-mr/pbs/brussels_capital.pdf
6 This section is based on Robert Geyer's award winning essay for the *European Voice* newspaper, 19 September 2001.

7 Development

1 For a broader discussion of complexity and development see Rihani (2002).
2 Both figures were created by Steve Pickering, www.stevepickering.net. Please note, that in order to avoid data overlap, the figures for state expenditure as a percentage of GDP

were averaged for the 1971–2006 period creating the impression that the relationship does not change over time. Interestingly, in the case of the advanced industrial countries, the relationship was generally remarkably stable. Unsurprisingly, for the least successful economies it often fluctuated wildly. Also note, in order to maintain the legibility of the figures the number of countries were reduced in the second figure. More generally, it is important to re-emphasise that our simplified fitness landscapes are not fitness landscapes as defined by Coveney and Highfield (1995). In their definition, the fitness landscapes examines the interrelationship between the factors influencing the fitness of a particular actor or set of actors in a given system. In our simplified fitness landscapes, though we use a three-dimensional structure, there is no direct interrelationship between the countries and their economic performance. Hence, we are unable to examine the interrelationships that create this particular three-dimensional outcome. However, as a metaphorical tool for moving away from traditional linear interpretations we believe that this simplified fitness landscape is valid and conforms to a general complexity framework.
3 See http://www.un-documents.net/ocf-01.htm#II

8 Planning dreams into nightmares

1 Speech, 21 December 2006
2 In June 2009, the British government agreed to set up an inquiry 'learn the lessons' of the war. However, it was to be held in secret and would not apportion blame.
3 See Barnett (2004), Marshall (2007), Packer (2005), Phillips (2005), Ricks (2006) and Woodward (2006).
4 Richard Cheney, Secretary of Defence, 'The Difficulties Encountered in the Gulf War Crisis', The Gulf War: A First Assessment, Soref Symposium 1991, Washington Institute for Near East Policy. Available at: http://www.washingtoninstitute.org/templateC07.php?CID=55
5 *Los Angeles Times*, 'Bush was twice warned of Iraq challenges', 26 May 2007. *Washington Post*, 'Analysts Warnings of Iraq Chaos, 26 May 2007.
6 See http://intelligence.senate.gov/prewar.pdf
7 Newspapers were full of these after-thoughts. Geoff Hoon (British defence secretary during the invasion) revealed that the British disagreed with the US over two key decisions in May 2003 (*Guardian*, 2 May 2007). Sir Mike Jackson, who headed the British Army in the 2003 war depicted the US as 'intellectually bankrupt' (*Guardian* 1 September 2007).
8 *New York Times*, 'Iraq Reconstruction Is Doomed', 3 May 2007.
9 *New York Times*, 'Former Top Genral in Iraq Faults Bush Adminstartion', 12 October 2007.
10 *Los Angeles Times*, 'War and peace, the Army way', 28 February 2008.
11 See http://www.brookings.edu/saban/iraq-index.aspx
12 At the annual meeting of the American Psychiatric Association on 5 May 2008, a top researcher reported that suicides among veterans of the wars in Iraq and Afghanistan may exceed the combat death toll.
13 This was recorded by UNICEF (*State of the World's Minorities 2007*, Minority Rights Group International, 2007: 107).
14 See: http://www.refugees.org/uploadedFiles/Investigate/Publications_&_Archives/WRS_Archives/2007/SilentSurge.pdf
15 See: www.globalcomplexity.org/PlightofIraqiacademics.htm
16 As the stupidity of this order became increasingly clear, Bremer suggested that it wasn't his order but came from Rumsfeld's office (Bremer and McConnell 2006: 39).
17 Ashby's Law of Requisite Variety: to control a system the controller must have sufficient variety and flexibility to match the variety presented by a system (Beer 1959).

9 Exploding the myths of terrorism

1 For these quotations and others see Abbott *et al.* (2007).
2 It is important to point out that viewing terrorism as a complex phenomenon has been around for some time (see Ranelagh 1992).
3 A project which the Obama administration has been working hard to distance itself from, but cannot avoid its ramifications.
4 *Domestic Violence: A National Plan*, UK Home Office, March 2005.
5 See http://www.homeoffice.gov.uk/crime-victims/reducing-crime/domestic-violence/

10 Conclusion

1 And, we would argue, this transformation is becoming increasingly apparent in intellectual and policy arenas. See the excellent work of Haynes (2003), Seddon (2008) and, of course, Ralph Stacey as well as *Tackling Wicked Problems*, published by the Australian Public Service Commission (2007), available on www.apsc.gov.au/publication07/wickedproblems.pdf
2 As of mid-2009 the Brown government has been making a number of U-turns in education policy that mirror those in health policy. Due to growing financial constraints, targets and testing have been reduced and localities and local actors are supposed to be given more control and responsibility. Clearly for the centre, at times of fiscal cutbacks and retrenchment it is much better to have the locals making the unpopular decisions. The tragedy of this strategy is that once finances return to normal recentralisation occurs and the process goes around again.

Bibliography

Abbott, C, Rogers, P. and Sloboda, J. (2007) *Beyond Terror: The Truth about the Real Threats to Our World*, London: Rider and Co.
Adler, E. (1997) 'Seizing the Middle Ground: Constructivism in World Politics', *European Journal of International Relations*, vol. 3, no. 3, 319–63.
Allen, P. (2001) 'What is Complexity Science? Knowledge of the Limits to Knowledge', *Emergence*, vol. 3, no. 1, 24–42.
Allison, G. (1971) *The Essence of Decision: Explaining the Cuban Missile Crisis*, Boston: Little Brown.
Anderson, L. and Stanfield, G. (2004) *The Future of Iraq: Dictatorship, Democracy or Division*, London: Palgrave
Ashley, R. (1986) 'The Poverty of Neorealism', in Keohane, R. (ed.) *Neorealism and Its Critics*, New York: Columbia University Press.
Australian Government Public Service Commission (2007) *Tackling Wicked Problems: A Public Policy Perspective*, Commonwealth of Australia.
Axelrod, R. (1997) *The Complexity of Cooperation: Agent Based Models of Competition and Collaboration*, New Jersey: Princeton University Press.
Axelrod, R. and Cohen, M. D. (2000) *Harnessing Complexity: Organisational Implications of Scientific Frontier*, New York: Free Press.
Bache, I. and Flinders, M. (eds.) (2005) *Multi-Level Governance*, Oxford: Oxford University Press.
Bache, I. and Jordon, A. (eds.) (2006) *The Europeanization of British Politics*, London: Palgrave.
Bak, P. (1997) *How Nature Works: The Science of Self-Organized Criticality*, Oxford: Oxford University Press.
Barber, B. (1984) *Strong Democracy: Participatory Politics for a New Age*, Berkeley: University of California Press.
Barnett, T. (2004) *The Pentagon's New Map: War and Peace in the Twenty-First Century*, New York: G.P. Putnams and Sons.
Barnett, W., Geweke, J. and Shell, K. (eds.) (1989) *Economic Complexity*, Cambridge: Cambridge University Press.
Barry, A. and Walters, W. (2003) 'From Euratom to complex systems', *Alternatives: Global, Local, Political*, vol. 28, no. 2, 305–29.
Bar-Yam, Y. (1997) *Dynamics of Complex Systems*, Reading: Perseus Press.
Batatu, H. (2004) *The Old Social Classes and the Revolutionary Movements of Iraq*, London: Saqi Books.

Beck, U. (1992) *The Risk Society: Towards a New Modernity*, London: Sage.

Beer, S. (1959) *Cybernetics and Management*, London: English University Press

Benton, T. (1999) 'Radical Politics – Neither Left nor Right', in M. O'Brien, S. Penna and C. Hay (eds.) *Theorising Modernity: Reflexivity, Environment and Identity in Giddens' Social Theory*, Longman: London.

Beinhocker, E. (2007) *The Origin of Wealth: Evolution, Complexity, and the Radical Remaking of Economics*, New York: Random House.

Bevir, M. (1999) *The Logic of the History of Ideas*, Cambridge: Cambridge University Press.

Bhaskar, R. (1986) *Scientific Realism and Human Emancipation*, London: Verso.

Blackman, T. (2006) *Placing Health: Neighbourhood Renewal, Health Improvement and Complexity*, Bristol: Policy Press.

Bremer, L.P. III and McConnell, M. *My Year in Iraq: The Struggle to Build a Future of Hope*, New York: Simon & Schuster, 2006.

Bogg, J. and Geyer, R. (eds.) (2007) *Complexity, Science and Society*, Oxford: Radcliffe Publishing.

Bonoli, G., George, V. and Taylor-Gooby, P. (2000) *European Welfare Futures: Towards a Theory of Retrenchment*, Cambridge: Polity Press.

Bottomore, T. (1993) *Elites and Society*, London: Routledge.

Brandt Commission (1980) *North–South: a programme for survival*, London: Pan Books.

Briggs, J. and Peat, F. D. (1990) *Turbulent Mirror: An Illustrated Guide to Chaos Theory and the Science of Wellness*, New York: Harper and Row.

Brown, L.R. (2001) *State of the world 2001*, London: Earthscan.

Brugha, R. and Zwi, A. (2002) 'Global Approaches to Private Sector Provision: Where is the Evidence' in Lee, K., Buse, K. and Fustukian, S. (eds.) *Health Policy in a Globalising World*, Cambridge: Cambridge University Press.

Burke, J. (2004) *Al'Qaeda: The True Story of Radical Islam*, London: Penguin.

Bush, G. and Scowcroft, B. (1998) *A World Transformed*, New York: Knopf.

Byrne, D. (1998) *Complexity Theory and the Social Sciences*, London: Routledge.

Capra, F. (1991) *The Tao of Physics*, Boston: Shambhala Publications.

—— (1996) *The Web of Life*, New York: Flamingo.

Cassis, Y. (1994) *Business Elites*, Cheltenham: Edward Elgar.

Casti, J. (1994) *Complexification: Explaining a Paradoxical World Through the Science of Surprise*, New York: Harper Collins.

Caufield, C. (1996) *Masters of illusion*, London: Macmillan.

Cederman, L. E. (1997) *Emergent Actors in World Politics*, Princeton: Princeton University Press.

Chapman, J. (2002) *System Failure*, Demos: London

Chandler, A.D., Amatori, F. and Hikino, T., (eds) (1997) *Big Business and the Wealth of Nations*, Cambridge: Cambridge University Press.

Chang, H. (2003) *Kicking Away the Ladder: Development Strategy in Historical Perspective*, London: Anthem Press.

Chang, H (2007) *Bad Samaritans: Rich Nations, Poor Policies and the Threat to the Developing World*, New York: Random House.

Checkel, J. (1998) 'The Constructivist Turn in International Relations Theory', *World Politics*, vol. 50, no. 2, 324–48.

—— (1999) 'Social Construction and Integration', *Journal of European Public Policy*, vol. 6, no. 4, 545–60.

Checkland, P. and Poulter, J. (2006) *Learning for Action: A Short Definitive Account of Soft Systems Methodology*, Chichester: John Wiley.
Childs, S. (1936) *Sweden: The Middle Way*, New York: Penguin Books.
Chomsky, N. (2006) *Failed States: The Abuse of Power and Assault on Democracy*, New York: Hamish Hamilton.
Christiansen, T., Jørgensen, K. E. and Wiener, A. (2001) *The Social Construction of Europe*, London: Sage.
Chryssochoou, D. (2001) *Theorizing European Integration*, London: Sage.
Cilliers, P. (1998) *Complexity and Postmodernism: Understanding Complex Systems*, London: Routledge.
Cioffi-Revilla, C. (1998) *Politics and Uncertainty*, Cambridge: Cambridge University Press.
Collier, P. (1998) 'The Political Economy of Ethnicity', *Annual World Bank Conference on Development Economics*, 20–21 April 1998, York: International Action Centre.
Cooper, H. and Geyer, R. (2007) *Riding the Diabetes Rollercoaster*, Oxford: Radcliffe Publishing.
Cooper, H. and Geyer, R. (2008) 'Using complexity for improving educational research in health care', *Journal of Social Science and Medicine*, Vol. 67, no. 1, 177–82.
Cooper, H. and Geyer, R. (2009) 'What can Complexity Do for Diabetes Management: Linking Theory to Practice', *Journal of Evaluation in Clinical Practice*, 15, 4.
Cooper H, Braye, S and Geyer, R (2004) 'Complexity and interprofessional education', *Learning and Teaching in Health and Social Care*, vol. 3, no. 4, 179–89.
Coveney, P. and Highfield, R. (1990) *The Arrow of Time*, London: Harper Collins.
—— (1995) *Frontiers of Complexity: The Search for Order in a Chaotic World*, London: Faber and Faber.
Cowan, G. A., Pines, D. and Metzger, D. (eds.) (1994) *Complexity: Metaphors, Models and Reality*, CA: Addison Wesley.
Cram, L. (1997) *Policy-making in the European Union: Conceptual Lenses and the Integration Process*, London: Routledge.
Currie, C. J., Kraus, D. and Morgan, C.L. et al. (1997) 'NHS acute sector expenditure for diabetes: the present, future, and excess in-patient cost of care', *Diabetic Medicine*, vol. 14, 686–92.
Day, R. and Samuelson, P. (1994) *Complex Economic Dynamics*, Boston: The MIT Press.
Dahrendorf, R. (1999) 'Whatever Happened to Liberty?', *New Statesman*, 6 September, 25–27.
Davies, P. (1987) *Superforce: The Search for a Grand Unified Theory of Everything*. London: Unwin Hyman.
—— (1989) *The New Physics*, Cambridge: Cambridge University Press.
Dawkins, R. (1996) *The Blind Watchmaker: Why the Evidence of Evolution Reveals a Universe Without Design*, New York: W.W. Norton.
—— (2006) *The Selfish Gene: 30th Anniversary Edition*, Oxford: Oxford University Press.
Deacon, B. (1997) *Global Social Policy: International Organizations and the Future of Welfare*, London: Sage.
Delanty, G. (1997) *Social Science: Beyond Constructivism and Realism*, Buckingham: Open University Press.
De Soto, H. (2000) The Mystery of Capital: *Why Capitalism Triumphs in the West and Fails Everywhere Else*, London: Bantam Press.
Dewey, J. (1929) *The Quest for Certainty: A Study of the Relation of Knowledge and Action*, New York: Minton-Balch.

Diez, T. (1999) 'Speaking "Europe": The Politics of Integration Discourse', *Journal of European Public Policy*, vol. 6, no. 4, 598–613.

Driver, S. and Martell, L. (2000) 'Left, Right and the Third Way', *Policy and Politics*, vol. 28, no. 2.

Downs, A. (1957) *An Economic Theory of Democracy*, New York: Harper.

Dunn, M. (2007) 'Securing the digital age: the challenges of complexity for critical infrastructure protection and international relations theory', in Eriksson, J. and Giacomello, G. (eds.) *International Relations and Security in the Digital Age*, London: Routledge, 85–105.

Eagleton, T. (1996) *The Illusions of Postmodernism*, Oxford: Blackwell.

Easterly, W. (2001) *The Elusive Quest for Growth: Adventures and Misadventures in the Tropics*, Cambridge: MIT Press.

Einhorn, E. and Logue, J. (2003) *Modern Welfare States: Scandinavian Politics and Policy in a Global Age*, Westport CT, London: Praeger.

Elliott, E. and Kiel, L. (eds.) (1999) *Nonlinear Dynamics, Complexity and Public Policy*, New York: Nova Science Publishers.

Esping-Andersen, G. (1985) *Politics Against Markets: The Social Democratic Road to Power*, Princeton: Princeton University Press.

Esping-Andersen, G. (ed.) (1996) *Welfare States in Transition: National Adaptations in Global Economics*, London: Sage.

Etzioni-Halvey, E. (1997) *Classes and Elites in Democracy and Democratisation*, New York: Garland Publishing.

Evans, J.R., Hall, K.L. and Watford, J. (1981) 'Health care in the developing world', *New England Journal of Medicine*, vol. 305, no. 19, 1117–27.

Eve, R.A., Horsfall, S. and Lee, M. E. (eds.) (1997) *Chaos, Complexity and Sociology: Myths, Models and Theories*, London: Sage.

Evers, A. and Laville, J. L. (eds.) (2004) *The Third Sector in Europe*, Cheltenham: Edward Elgar.

Faux, J. (Spring 1999) 'Lost on the Third Way', *Dissent*, vol. 46, no. 2.

Feith, D. (2008) *War and Decision: Inside the Pentagon at the Dawn of the War on Terrorism*, London: HarperCollins.

Ferguson, M. (1983) *The Aquarian Conspiracy*, London: Paladin.

Ferrera, M. and Rhodes, M. (eds.) (2000) 'Recasting European Welfare States', *West European Politics*, vol. 23, no. 2.

Fraser, S. W. and Greenhalgh, T. (2001) 'Complexity science: coping with complexity: educating for capability', *British Medical Journal*, 323, 799–803.

Fukuyama, F. (1992) *The End of History and the Last Man*, London: Penguin.

Gaddis, J. L. (2002) *The Landscape of History: How Historians Map the Past*, Oxford: Oxford University Press.

Gatrell, A. (2005) 'Complexity theory and geographies of health: a critical assessment', *Social Science and Medicine*, vol. 60, no. 12, 2661–71.

Gell-Mann, M. (1994) *The Quark and the Jaguar*, Boston MA: Little Brown.

George, S. (1994) *A Fate Worse Than Debt*, London: Penguin.

Georgescu-Roegen, N. (1999) *The Entropy Law and the Economic Process*. Cambridge, Massachusetts: MIT Press.

Geyer, R. (1997) *The Uncertain Union: British and Norwegian Social Democrats in an Integrating Europe*, Aldershot: Avebury.

—— (2000) *Exploring European Social Policy*, Cambridge: Polity Press.

—— (2001) 'Can EU Social NGOs Promote EU Social Policy?' *Journal of Social Policy*, 30, 3, 2001.
—— (2003) 'Beyond the Third Way: The Science of Complexity and the Politics of Choice', *British Journal of Politics and International Relations*, vol. 5, no. 2, 237–57.
—— (2003) 'European Integration, the Problem of Complexity and the Revision of Theory', *Journal of Common Market Studies*, vol. 41, no. 1, 15–35.
Geyer, R., Ingebritsen, C. and Moses, J. (eds.) (2000) *Globalization, Europeanization and the End of Scandinavian Social Democracy?*, London: Macmillan.
Geyer, R. and MacIntosh (2005) *Integrating UK and European Social Policy: The Complexity of Europeanisation*, Oxford: Radcliffe Publishing, 2005.
Giddens, A. (1994) *Beyond Left and Right: The Future of Radical Politics*, Cambridge: Polity Press.
—— (1998) *The Third Way: The Renewal of Social Democracy*, Cambridge: Polity Press.
—— (2000) *The Third Way and Its Critics*, Cambridge: Polity Press.
Gladston, I. (1981) *Social and Historical Foundations of Modern Medicine*, New York: Brunner/Mazel Publishers.
Gladwell, M. (2001) *The Tipping Point: How Little Things Can Make a Big Difference*, London: Abacus.
Glasby, J., Peck, E., Ham, C. and Dickinson, H. (April 2007) 'Things can only get better?-the argument for NHS independence', Health Services Management Centre, School of Public Policy, University of Birmingham. http://www.newscentre.bham.ac.uk/documents/Things_can_only_get_better–the_argument_for_NHS_independence.pdf
Gleick, J. (1987) *Chaos*, London: Sphere.
Goodwin, B. (1994) *How the Leopard Changed its Spots: The Evolution of Complexity*, New York: Scribners.
Gorsky, M. (2008) 'The British National Health Service 1948–2008: A Review of the Historiography', *Social History of Medicine*, vol. 21, no. 3, 437–60.
Gould, S. J. (1981) *The Mismeasure of Man*, New York: Norton.
—— (1989) *Wonderful Life: The Burgess Shale and the Nature of History*, New York: Norton.
Gourevitch, P. (1986) *Politics in Hard Times: Comparative Responses to International Economic Crises*, Ithaca: Cornell University Press.
Grande, E. and Pauly, L. (2005) *Complex Sovereignty: Reconstituting Political Authority in the Twenty-First Century*, Toronto: Toronto University Press.
Green, D. and Shapiro, I. (1996) *Pathologies of Rational Choice Theory: A Critique of Applications in Political Science*, New Haven: Yale University Press.
Gulbenkian Commission (1996) *Open the Social Sciences*, Stanford: Stanford University Press.
Hobsbawm, E. (1994) *Age of Extremes: The Short Twentieth Century*, London: Michael Joseph.
Haggis, T. (2007) 'Conceptualising the Case in Adult and Higher Education Research: a Dynamic Systems View' in Bogg, J. and Geyer, R. *Complexity, Science and Society*, Oxford: Radcliffe Publishing.
Hague, R and Harrop, M. (2004) *Comparative Government and Politics*, London: Palgrave.
Hall, S. (1998) 'The Great Moving Nowhere Show', *Marxism Today*, November/December, 9–14.
Hanf, K. and Soetendorp, B. (eds.) (1998) *Adapting to European Integration: Small States and the European Union*, London: Longman.

Harris, G. (2007) *Seeking Sustainability in an Age of Complexity*, Cambridge: Cambridge University Press.

Harrison, N. E. (ed) (2006) *Complexity in World Politics: Concepts and Methods of a New Paradigm*, New York: State University of New York Press.

Hawking, S. (1988) *A Brief History of Time*, London: Bantam Press.

Hay, C. (2001) 'Globalization, Economic Change and the Welfare State: The Vexatious Inquisition of Taxation?', in Sykes, R. et al. (eds.) *Globalization and European Welfare States: Challenges and Change*, London: Palgrave.

—— (2002) *Political Analysis: A Critical Introduction*, Basingstoke: Palgrave.

Hayek, F. A. (1967) *Studies in Philosophy, Politics and Economics*, Chicago: University of Chicago Press.

Haynes, P. (2003) *Managing Complexity in the Public Services*, Maidenhead: Open University Press.

Held, D. and McGrew, A. (2007) *Globalization/Anti-Globalization*, Cambridge: Polity Press.

Herman, E. S. and Chomsky, N. (1994) *Manufacturing Consent*, Montreal: Black Rose Books.

Herrmann, H. (1998) *From Biology to Sociopolitics: Conceptual Continuity in Complex Systems*, New Haven: Yale University Press.

Heikal, M. (1992) *Illusions of Triumph*, London: HarperCollins.

Hirst, P. and Thompson, G. (1996) *Globalization in Question*, Cambridge: Polity Press.

Hodgson, G. (1997) *Economics and Evolution*, Ann Arbor: University of Michigan Press.

Hodson, D. and Maher, I. (2001) 'The Open Method as a New Mode of Governance', *Journal of Common Market Studies*, vol. 39, no. 4, 719–46.

Holland, J. and Langton, C. G. (1992) 'Life at the edge of chaos' in Langton, C.G., Taylor, C. Farmer, J. D. and Rasmussen, S. (eds.) *Artificial Life II*, New York: Addison-Wesley.

Holt, T. (ed.) (2004) *Complexity for Clinicians*, Oxford: Radcliffe Publishing.

Hood, C. (1991) 'A Public Management for All Seasons?', *Public Administration* vol. 69, no. 1, 3–19

Hooghe, L. and Marks, G. (2001) *Multi-level Governance and European Integration*, Rowman and Littlefield: Lanham.

Hoogvelt, A. (2001) *Globalisation and the Postcolonial World*, Basingstoke: Palgrave.

Horgan, J. (1996) *The End of Science: Facing the Limits of Knowledge in the Twilight of the Scientific Age*, New York: Broadway Books.

Horowitz, D. L. (1998) '*Structure and Strategy in Ethnic Conflict: A Few Steps Toward Synthesis*', Annual World Bank Conference on Development Economics, in Pleskovic, B. and Stiglitz, J. (eds.) (1999), pp. 345–70, Washington D.C.: World Bank.

Hourani, A (1991) *A History of the Arab Peoples*,

Jervis, R. (1998) *System Effects*, Princeton: Princeton University Press.

Johnson, S. (2002) *Emergence: The Connected Lives of Ants, Brains, Cities, and Software*, New York: Touchstone.

Jørgensen, K. E. (ed.) (1997) *Reflective Approaches to European Governance*, Basingstoke: Macmillan.

Katzenstein, P. (1985) *Small States in World Markets: Industrial Policy in Europe*, Ithaca: Cornell University Press.

Katzenstein, P., Keohane, R. and Krasner, S. (1998) International Organization and the Study of World Politics, *International Organization*, vol. 52, no. 4, 645–85.

Kavalski, E. (2007) 'The fifth debate and the emergence of complex international relations theory: notes on the application of complexity theory to the study of international life', *Cambridge Review of International Affairs*, vol. 20, no. 3.
Kauffman, S. (1993) *The Origins of Order*, Oxford: Oxford University Press.
—— (1995) *At Home in the Universe*, London: Viking.
—— (1996) *At Home in the Universe: The Search for the Laws of Self-Organisation and Complexity*, Oxford: Oxford University Press.
Kendall, J. (2003) *The Voluntary Sector: Comparative Perspectives on the UK*, London: Routledge.
Kendall, J. and Anheier, H. (eds.) (2001) *Third Sector Policy at the Crossroads: An International Nonprofit Analysis*, London: Routledge.
Kennedy, P. M. (1989) *The Rise and Fall of Great Powers: Economic Change and Military Conflict from 1500 to 2000*, New York: Random House.
Keohane, R. and Nye, J. (1997) *Power and Interdependence: World Politics in Transition*, Boston: Little Brown.
Kernick, D. (ed.) (2004) *Complexity and Healthcare Organization: A View from the Street*, Oxford: Radcliffe.
Kiel, L.D. and Elliot, E. (eds.) (1997) *Chaos Theory in the Social Sciences: Foundation and Application*, Ann Arbor: University of Michigan Press.
Kiely, R. and Marfleet, P. (1998) *Globalisation and the Third World*, London: Routledge.
King, I. (2000) *Social Sciences and Complexity: The Scientific Foundations*, Huntington NY: Nova Science Publishers.
Klein, N. (2007) *The Shock Doctrine*, New York, Penguin.
Knutsen, T. (1997) *A History of International Relations Theory* 2nd edition, Manchester: Manchester University Press.
Kondepudi, D. and Prigogine, I. (1998) *Modern Thermodynamics: From Heat Engines to Dissipative Structures*, New York: Wiley.
Korpi, W. (1983) *The Democratic Class Struggle*, London: Routledge and Kegan Paul.
Krasner, S. (1982) 'American Policy and Global Economic Stability' in Avery and Rankin, (eds.) *America in a Changing World Economy*, New York: Longman.
Krasner, S. D. (ed.) (1983) *International Regimes*, Ithaca: Cornell University Press.
Kravtsov, Y. A. and Kadtke, J. B. (eds.) (1996) *Predictability of Complex Dynamical Systems*. Berlin, Springer-Verlag
Kuhn, T. (1962) *The Structure of Scientific Revolutions*, Chicago: University of Chicago Press.
Kuhnle, S. (2000) *Survival of the European Welfare State*, London: Routledge.
Lafointaine, O. (1998) 'The Future of German Social Democracy', *New Left Review*, no. 227, 72–87.
Lake, D. A. (1991) 'British and American Hegemony Compared', in M. Fry, (ed.) *History, The White House and the Kremlin*, London: Pinter.
Langton, C.G., Taylor, C. Farmer, J. D. and Rasmussen, S. (eds.) (1992) *Artificial Life II*, New York: Addison-Wesley.
Lasch, C. (1995) *The Revolt of the Elites and the Betrayal of Democracy*, New York: W. W. Norton.
Leibfred, S. and Pierson, P. (eds.) (1995) *European Social Policy: Between Fragmentation and Integration*, Washington DC: Brookings Institute.
Lerner, R., Nagai, A. K. and Rothman, S. (1996) *American Elites*, New Haven CT: Yale University Press.

Lesmoir-Gorson, N., Rood, W. and Edney, R. (2000) *Introducing Fractal Geometry*, Cambridge: Icon Books.
Levy, J. (1999) 'Vice into Virtue? Progressive Politics and Welfare Reform in Continental Europe', *Politics and Society*, vol. 27, no. 2, 239–73.
Lewin, R. (1999) *Complexity: Life at the Edge of Chaos* 2nd edition, Chicago: University of Chicago Press.
Lightfoot, S. (1999) 'Prospects for Euro-socialism', *Renewal*, vol. 7, no. 2, 7–17.
Lindblom, C. (1959) 'The Science of "Muddling Through"', *Public Administration Review*, vol. 19, no. 2, 79–88.
Lorenz, E. (1993) *The Essence of Chaos*, Seattle: University of Washington Press.
Lovelock, J. E. (1979) *Gaia: A New Look at Life on Earth*, Oxford: Oxford University Press.
Lynn, J. and Jay, A. (1989) *The Complete Yes Prime Minister*, London, BBC books.
Lyotard, J. F. (1984) *The Postmodern Condition: A Report on Knowledge*, Minneapolis: University of Minnesota Press.
Maddison, A. (1982) *Phases of Capitalist Development*, Oxford: Oxford University Press.
—— (2006) *The World Economy*, New York: OECD Development Centre Studies.
—— (2007) *Contours of the World Economy*. Oxford: Oxford University Press.
Mainzer, K. (1997) *Thinking in Complexity: The Complex Dynamics of Matter, Mind, and Mankind*, Berlin: Springer.
Marger, M. (1986) *Elites and Masses: An Introduction to Political Sociology*. London: Wadsworth Publishing.
Margolis, J. (1993) *The Flux of History and the Flux of Science*, Berkeley: University of California Press.
Marks, G., Hooghe, L. and Blank, K. (1996) 'European Integration from the 1980s: State-centric v. Multi-level Governance', *Journal of Common Market Studies*, vol. 34, no. 5, 342–78.
Marks, G, Nielsen, F., Ray, L. and Salk, J. (1996) 'Competencies, Cracks and Conflicts: Regional Mobilization in the European Union', in G. Marks, F. Scharpf, P. Schmitter, ? and Streeck, W. (eds.) *Governance in the European Union*, London: Sage.
McLean, I. (ed.) (1996) *The Oxford Concise Dictionary of Politics*, Oxford: Oxford Paperbacks.
Martin, M. and McIntyre, L. (eds) (1994) *Readings in the Philosophy of Social Science.* Cambridge, Massachusetts: MIT Press.
Meisler, S. (1995) 'Dateline UN: A New Hammarskjold?' *Foreign Policy*, no. 98, Spring 1995.
Michels, R. (1915) *Political Parties: A Sociological Study of the Oligarchic Tendencies of Modern Democracy*, Glencoe, Illinois: Free Press.
Miles, L. (2005) *Fusing with Europe: Sweden in the European Union*, London: Ashgate.
Mirowski, P. (1994) *Natural Images in Economic Thought*, Cambridge: Cambridge University Press.
Moravcsik, A. (1993) 'Preferences and Power in the European Community: A Liberal Intergovernmentalist Approach', *Journal of Common Market Studies*, vol. 31, no. 4, 473–524.
—— (2001) 'Constructivism and European Integration: A Critique'. In T. Christiansen *et al.* (eds.) *The Social Construction of Europe* (London: Sage).
Morowitz, H. (2002) *The Emergence of Everything: How the World Became Complex*, Oxford: Oxford University Press.

Nicolis, G. and Prigogine, I. (1989) *Exploring Complexity*, New York: W.H. Freeman.
Ohmae, K. (1994), *The Borderless World: Power and Strategy in the Interlinked Economy*, London: HarperCollins, 1994.
Ormerod, P. (1994) *The Death of Economics*, London: Faber and Faber.
—— (1998) *Butterfly Economics*, London: Faber and Faber.
Onuf, N. (1989) *A World of our Making: Rules and Rule in Social Theory and International Relations*, Columbia: University of South Carolina Press.
Osborne, S. (2008) *The Third Sector in Europe: Continuity and Change*, London: Routledge.
Oxman, A. D., Sackett, D. L., Chalmers, I., *et al* (2005) 'A Surrealistic Mega-analysis of Redisorganisation Theories. *Journal of the Royal Society of Medicine*, 98, 563–68.
Packer, G. (2005) *The Assassins' Gate: America in Iraq*, New York: Farrar Straus Giroux.
Peat, F. D. (1991) *Superstrings and the Search for a Theory of Everything*, London: Sphere Books.
Phillips, D. (2005) *Losing Iraq*, New York: Westview Press.
Pierson, P. (1994) *Dismantling the Welfare State? Reagan, Thatcher, and the Politics of Retrenchment*, Cambridge: Cambridge University Press.
Player, S. and Leys, C. (2008) *Confuse and Conceal: The NHS and Independent Sector Treatment Centres*, Monmouth, Wales: Merlin Press.
Plesk, P. E. and Greenhalgh, T. (2001) 'The challenge of complexity in health care', *British Medical Journal*, 323, 625–28.
Plesk, P, and Wilson, T. (2001) 'Complexity Science: complexity, leadership, and management in healthcare organisations', *British Medical Journal*, 323, 746–49.
Pollack, M. (2001) 'International Relations Theory and European Integration', *Journal of Common Market Studies*, vol. 39, no. 2, 221–44.
Poincaré, H. (2001) *The Value of Science: The Essential Writings of Henri Poincaré*, New York: The Modern Library.
Power, M. (1997) *The Audit Society: Rituals of Verification*, Oxford: Oxford University Press.
Prigogine, I. and Stengers, I. (1984) *Order Out of Chaos: Man's New Dialogue with Nature*, New York: Bantam Books.
Putnam, R (2001) *Bowling Alone: The Collapse and Revival of American Community*, New York: Simon and Schuster.
Ranelagh (1992: 225), *CIA: A History*, BBC Books
Rasch, W. and Wolfe, C. (eds.) (2000) *Observing Complexity: Systems Theory and Postmodernity*, Minneapolis: University of Minnesota Press.
Reiser, S. J. (1978) *Medicine and the Reign of Technology*, Cambridge: Cambridge University Press.
Rescher, N. (1998) *Complexity: A Philosophical Overview*, London: Transaction Publishers.
Rengger, N. J. (2000) *International Relations, Political Theory and the Problem of Order*, London: Routledge.
Rhodes, M. (1996) 'Globalization, Labour Markets and Welfare States: a Future of Competitive Corporatism',. *Journal of European Social Policy*, vol. 6, no. 4, 305–27.
Rich, B. (1994) *Mortgaging the Earth*, Boston MA: Beacon Press.
Richards, D. (ed.) (2000) *Political Complexity*. Ann Arbor, Michigan: University of Michigan Press.
Richards, D. and Smith, M. (2002) *Governance and Public Policy in the UK*, Oxford: Oxford University Press.

Richardson, K and Cilliers, P. (2001) 'What is Complexity Science? A View from Different Directions', *Emergence*, vol. 3, no. 1, 5–24.
—— (2007) *Explorations in Complexity Thinking*, Mahwah, New Jersey: ISCE Publishing.
Ricks, T. E. (2006) *Fiasco*, London: Allen Lane.
Rihani, S. and Geyer, R. (2001) Complexity: an appropriate framework for development? *Progress in Development Studies*, vol. 1 no. 3, 237–45.
Rihani, S. (2002) *Complex Systems Theory and Development Practice: Understanding Non-linear Realities*, London: Zed Books.
Rihani, S. (2010) 'English National Health Service: a complex system at the mercy of central planning', *Emergence*, Winter/Spring.
Rorty, R. (1991) Objectivism, Relativism and Truth, Cambridge: Cambridge University Press.
Rosamond, B. (2000) *Theories of European Integration*, London: Macmillan.
Rose, N. (1999) 'Inventiveness in Politics', *Economy and Society*, vol. 28, no. 3, 467–93.
Rostow, W. (1960) *The Stages of Economic Growth: A Non-Communist Manifesto*, Cambridge: Cambridge University Press.
Rouse, J. and Smith, G. (1999) 'Accountability', in M. Powell (ed.) *New labour, New Welfare State? The 'Third Way' in British Social Policy*, Bristol: The Policy Press.
Ryan, A. (1999) 'Britain: Recycling the Third Way', *Dissent*. vol. 46, no. 2, 77–80.
Rycroft, R. and Kash, D. (1999) *The Complexity Challenge: Technological Innovation for the 21st Century*, London: Pinter.
Sandholtz, W. (1996) 'Membership Matters: Limits of the Functional Approach to European Institutions,' *Journal of Common Market Studies*, vol. 34, 403–29.
Samuelson, P.A. and Nordhaus, W.D. (1995) *Economics*, New York: McGraw-Hill.
Savit, Robert, 'Chaos on the trading floor' in Hall, N. (ed.) *Exploring Chaos: A Guide to the New Science of Disorder*, Norton: New York, 1991.
Sawyer, R.K. (2005) *Social Emergence: Societies as Complex Systems*, Cambridge, Cambridge University Press.
Scholte, J. A. (2005) *Globalization: A Critical Introduction*, Basingstoke/New York: Palgrave.
Scott, J. (1998) *Seeing Like a State*, New Haven: Yale University Press.
Seddon, J. (2008) *Systems Thinking in the Public Sector*, Axminster: Triarchy Press.
Sen, A. (2000) *Development as Freedom*. London: Anchor Books.
Serra, N. and Stiglitz, J. eds. (2008) *The Washington Consensus Reconsidered*, Oxford: Oxford University Press.
Simmons, M. (2003) 'Complexity theory in the management of communicable diseases', *Journal of Hospital Infection*, vol. 52, 87–92.
Skogly, S. (2006) *Beyond National Borders: States' Human Rights Obligations in International Cooperation*, Oxford: Intersentia.
Smith, D. H. (1973) *Confucius*, London: Temple Smith.
Smith, S. (2001) 'Social Constructivisms and European Studies', in Christiansen, T. *et al.* (eds.) *The Social Construction of Europe*, London: Sage.
—— (2001) 'Reflectivist and Constructivist Approaches to International Theory', in Baylis, J. and Smith, S. (eds.) *The Globalization of World Politics: An Introduction to International Relations* (2nd Ed.), Oxford: Oxford University Press.
Smith, J. and Jenks, C. (2006) *Qualitative Complexity: Ecology, Cognitive Processes and the Re-emergence of Structures in post-humanist Social Theory*, London: Routledge.
Stacey, R. (1993) *Strategic Management and Organizational Dynamics*, London: Financial Times/Prentice Hall.

—— (1996) *Complexity and Creativity in Organizations*, London: Berret-Koehler.
—— (1999) *Strategic Management and Organisational Dynamics: The Challenge of Complexity*, London: Financial Times/Prentice Hall.
—— (2000) *Strategic Management and Organisational Dynamics*, Harlow: Pearson Education.
Stacey, R., Griffin, D. and Shaw, P. (2000) *Complexity and Management: Fad or Radical Challenge to Systems Thinking?*, London: Routledge.
Stewart, I. (1989) *Does God Play Dice? The Mathematics of Chaos*, Oxford: Blackwell.
Steinberg, D. (2005) *Complexity in Healthcare and the Language of Consultation*, Oxford: Radcliffe Publishing.
Stiglitz, J. (1998) *Towards a New Paradigm for Development: Strategies, Policies, and Processes*, Prebisch Lecture, United Nation Conference on Trade and Development, 19 October 1998.
—— (2003) *Globalisation and its Discontents*, London: Penguin
Stiglitz, J. and Blimes, L. (2008) *The Three Trillion Dollar War: The True Cost of the Iraq Conflict*, London: Allen Lane
Streeck, W. and Schmitter, P. C. (1991) 'From National Corporatism to Transnational Pluralism: Organised Interests in the Single European Market,' *Politics and Society*, vol. 19, no. 2, 133–64.
Swank, D. (1998) 'Funding the Welfare State: Globalisation and the Taxation of Business in Advanced Market Economies', *Political Studies*, vol. 46, no. 4, 671–92.
Sweeney, K. and Griffiths, F. (eds.) (2002) *Complexity and Healthcare*, Oxford: Radcliffe Publishing.
Sweeney, K. (2006) *Complexity and Primary Care*, Oxford: Radcliffe Publishing.
Sykes, R., Palier, B. and Prior, P. (eds.) (2001) *Globalization and European Welfare States: Challenges and Change*, London: Palgrave.
Taylor, P. (1983) *The Limits of European Integration*, London: Croom Helm.
—— (2005) *Unruly Complexity: Ecology, Interpretation, Engagement*, Chicago: Chicago University Press.
Thrift, N. (1999) 'The Place of Complexity', *Theory, Culture and Society*, vol. 16, no. 3, 31–69.
Timberlake, L. (1991) *Africa in Crisis: The Causes, the Cures of Environmental Bankruptcy*, London: Earthscan.
Tosey, P. (2002) 'Teaching on the Edge of Chaos: Complexity Theory, Learning Systems and Enhancement', Learning and Teaching Support Network, Generic Centre.
Toye, J. (1987) *Dilemmas of Development: Reflections on the Counter-Revolution in Development Theory and Policy*, Oxford: Blackwell.
Tranholm-Mikkelsen, J. (1991) 'Neo-Functionalism: Obstinate or Obsolete? A Reappraisal in the Light of the New Dynamism of the EC', *Millennium*, vol. 20, no. 1, 1–22.
United Nations Development Programme (1999) *Human Development Report 1999*, New York: Oxford University Press.
—— (2002) *Human Development Report 2002*, New York: Oxford University Press.
UNICEF (1995) United Nations Children's Fund, *The State of the World's Children 1995*, New York: Oxford University Press.
—— (1996) United Nations Children's Fund, *The State of the World's Children 1995*, New York: Oxford University Press.
—— (1997) *The Progress of Nations*, New York: Oxford University Press.
—— (1998) United Nations Children's Fund, *The State of the World's Children 1998*, New York: Oxford University Press.

—— (1999) United Nations Children's Fund, *Trends in Child Mortality in the Developed World 1960–1996*, New York: UNICEF.
—— (2007) United Nations Children's Fund, *The State of the World's Minorities 2007*, New York: UNICEF.
Urry, J. (2003) *Global Complexity*, Cambridge: Policy Press.
Walby, S. (2007) *Globalization and Complex Inequalities*, London: Sage.
Waldrop, M. (1992) *Complexity: The Emerging Science at the Edge of Order and Chaos*, New York: Simon and Schuster.
—— (1994) *Complexity: The Emerging Science at the Edge of Order and Chaos*, London: Penguin Books.
Walker, R. B. J. (1993) *Inside/Outside: International Relations as Political Theory*, Cambridge: Cambridge University Press.
Wallerstein, I. (1983) *Historical Capitalism*, London: Verso.
Walsh, K. (1995) *Public Services and Market Mechanisms: Competition, Contracting and the New Public Management*, London: Macmillan.
Walshe, K. (2003) *Regulating Healthcare: A Prescription for Improvement?* London: Open University Press.
Wendt, A. (1992) 'Anarchy is What States Make of It: The Social Construction of Power Politics', *International Organization*, vol. 46, no. 2, 391–407.
—— (1999) *Social Theory of International Politics*, Cambridge: Cambridge University Press.
Wild, S., Roglic, G., Green, A., Sicree, R. and King, H. (2004) 'Global Prevalence of Diabetes: Estimates for the year 2000 and projections for 2030', *Diabetes Care*, vol. 27, 1047–53.
Wilson, B. (2001) *Soft Systems Methodology: Conceptual Model Building and its Contribution*, Chichester: John Wiley.
Wilson, E. O. (1998) *Consilience: The Unity of Knowledge*, New York: Vintage Books.
Woodward, B. (2002) *Bush at War: Inside the Bush Whitehouse*, New York: Simon and Schuster.
—— (2006) *State of Denial: Bush at War Pt.III*, New York: Simon and Schuster.
World Bank (2004) *World Development Report 2004*, New York: World Bank Publications.
World Commission on Environment and Development (1987) *Our Common Future*, Oxford: Oxford University Press.
World Health Organization (2000) *The World Health Report 2000*, Geneva: World Health Organisation.
Zunes, S. (2003) *Tinderbox*, London: Zed Book.

Index

Diagrams and tables are given in italics.

'A World Without Islam' (Fuller) 171
Adorno, Theodor 25
Afghanistan 141, 166–7, 169, 173
Al-Qaeda network 166, 173, 177, 179
Algeria 177
Allen, Peter 17
Allison, Graham 27, 59–60
Anderson, L. 199
Army Field Manual on Operations (US) 151
'arrow of time' 37, 54, *54*
Australian Government Public Service Commission 68–9
Aznar, José María 148

'balance of power' concept 111
Barber, Benjamin 76–7
Batatu, Hanna 157
Ba'th Party 158–61
Being and Nothingness (Sartre) 47
Beinhocker, E. 133
Bell, Michael 150
Berlusconi, Silvio 148
Beyond Left and Right (Giddens) 80
Billroth, Theodor 94
Bilmes, Linda 153
bin Laden, Osama 166–71, 181
biotic world: adaptation, survival, variety, 'good enough' 42–3; arrow of time 45; complexity in the EU 116; depth 45–6, 194n.16; evolution 43–4; explanation 18–20, *19*; frozen accidents 44–5; gateway events 45; key concepts of 41; punctuated equilibrium 44–5
Blackman, T. 133
Blair, Tony 148, 167
Blair Years, The (Campbell) 150

Bowling Alone (Putnam) 133–4
Bremer, Paul 158–9
Brown, Gordon 102, 169
Brundtland Report (WCED) 136
Building Britain's Future (2009) 102
Bush, George W. 148–9, 156–7, 167–8, 170–1, 173
'butterfly effects' 16, 44

Cameron, David 102
Campbell, Alastair 150
Canadian Institute of Governance 69
Castro, Fidel 96
Chalabi, Ahmad 155, 158
Chang, Ha-Joon 50, 130, 134
Chapman, Jake 100
Cheney, Richard 149, 155, 157
Childs, S. 122
Cilliers, P. 30
'Citizens as Partners' (OECD) 69
Clinton, Bill 74
Cohen, Michael 27
Collier, P. 135
common cold 189–91
communism 40, 194n.6
complex systems: the biotic world *see* biotic world; bounded freedom/diversity 47–8; elites 49; emergence 50–2; evolving societal framework 48–9; key concepts of conscious 46–7; the physical world 16–18, *18*; unpredictability 50–1
complexity: cascade of 53–6, *56*; and democracy 76–80; and the EU 123–4, *124*; fitness landscapes *see* fitness landscapes; and health *see* health; mapping of 57–60, *57*, *60*;

morality of 184–5; progressive/open 183–4; public policy foundations *32–3*; range of *59*; social science foundations *31–2*; soft systems methodology (SSM) 69–72, *71*; Stacey diagram 64–8, *65*, *67*; stakeholder involvement 68–71; the 'third sector' 85–8, 89–91; the 'third way' 80–4; the way forward 187–90; and the weak 185
Condorcet, Jean Antoine marquis de 20, 55
Confessions of an Economic Hit Man (Perkins) 135
Congressional Budget Office (US) 152–3
conscious systems 29, *29*
Consilience (Wilson) 13
constructivism 111–13
Coveney, Peter 18–20, 50
Cuba 95–6
'culture' 38

Dahrendorf, Ralf 83, 85
Darwin, Charles 43
Darzi, Lord 99–100, 101
De Soto, H. 136
Declaration of Independence (USA) 76
Definition of Terrorism, The (UK Government report) 166
Democratic Class Struggle (Korpi) 120
Descartes, René 9, 12
development policy: barriers to 134–6; case for radical change 130–1; conclusions 143–5; definition 10, 129; guiding a nation 138, *138–40*, 141–2, 196–7n.2; lopsided view of 132–3; realistic view of 136–7; self-organised complexity 133–4; slowness of framework change 142–3
Dewey, John 25, 27, 76
Dialectic of Enlightenment (Adorno/Horkheimer) 25
Downs, Antony 23
Duncan, William Henry 95–6

Eagleton, Terry 25–6
education 189–90, 198n.10
End of Science, The (Horgan) 13
Enlightenment, the 12
Esping-Andersen, Gosta 120–1
Essence of Decision, The (Allison) 59–60
European Union (EU): aid to Russia 56; capital for 122–3, 127–8, *127*; compatibility with third sector 88–90; complexity of 9–10, 114–18, *115*, 124–5, *124*, *125*; frozen accidents 44; and integration theory 112–14
Evans, J. R. 96
evolution 43

fitness landscapes 60–3, *62–3*, 80, 105–8, *108*, 138, *138–40*, 143, 196–7n.2
Foundation Trusts (FTs) 102
Freud, Sigmund 25
frozen accidents 44–6, 53–4, 78, 123–8
Fukuyama, Francis 85, 118, 123
Fuller, Graham 171
'Future of Iraq Project' (US State Department) 151

Gaia concept 19–20
'garbage can' model 27
gateway events 134
Gell-Mann, Murray 18
Giddens, Anthony 80–5
global warming 185
globalisation 9–10
Goethe, Johann Wolfgang von 25
Gorsky, M. 98
Gould, Stephen 45, 55
Gourevitch, Peter 120–1

Hall, K. L. 96
Hamilton, William 194n.12
Hay, Colin 26
Hayek, F. A. 25
Haynes, Paul 30–1
health: complex view of 94–5; diabetes 103–8; influence of order 92–7; introduction 92; National Health Service (English) 98–103; understanding of 109
'Health for All by the Year 2000' (UNICEF) 136
Heisenberg, Werner 15
Highfield, Roger 18–20, 50
Hirst, Paul 118
Hobbes, Thomas 20, 49
Hobsbawm, Eric 4–5
Hodson, D. 117
Horgan, John 13
Horkheimer, Max 25
Hospital Plan 1962 98
human beings 28–30
Hussein, Saddam 3, 149, 155, 159, 181
Huygens, Christiaan 14

incrementalism, theory of 27
Independent Sector Treatment Centres (ISTCs) 99

international arena: complexity map 114–15, *114*; European integration theory (EI) 112–18; Europeanisation 119, 122–3; globalisation 118–19; international relations (IR) 111, 113; introduction 110; Scandinavian exceptionalism 10, 119–22, 125–6, *126*
Iraq Index (report) 153
Iraq situation: the blame 151–2; and complexity 134–5, 137, 141, 156–63, *161*, 197n.16/17; the costs v benefits 152–4, 197n.12; expensive policy fiasco 146–7, 197n.2; introduction 10; predictability of failure 147–50, 197n.7; underlying questions 154–5
'iron cage' (Weber) 26
'iron law of oligarchy' (Michels) 27
'ironic science' 13

Kanbur, Ravi 142
Kant, Immanuel 25
Katzenstein, Peter 113, 120–2
Kauffman, S. 18, 133
Kavalski, E. 114
Kennedy, John F. 59–60, *60*
Keohane, R. 113
Ki-Moon, Ban 144
Korpi, Walter 120–1
Krasner, S. 113
Kuhn, Thomas 6, 114, 186, 192n.3

Laplace, Pierre Simon de 12–13
Law of Requisite Variety (Ashby) 197n.17
le Carré, John 170
Limón, Lavinia 153
linear paradigm *see* order, paradigm of
linear/non-linear phenomena 6
Liverpool 94–5
Lorenz, Edward 16, 39
Lorenzian Waterwheel 17
Lovelock, James 19–20
Lyotard, Jean-Francois 25

Maddison, Angus 130
Maher, I. 117
Mainzer, Klaus 46, 184
'manufactured risk/uncertainty' (Giddens) 80–1
March, James 27
Marx, Karl 123
Michels, Robert 27
Michelson, Albert 13
Millennium Development Goals Report 2008 (UN) 144

Milosevic, Slobodan 168
Modernisation Agency (DOH) 69
Moravcsik, Andrew 112
Mugabe, Robert 49
'multi-level governance' 27

'new managerialism' 83
New Public Management (NPM) 23–4
Newlands, James 95–6
Newton, Sir Isaac: and the Enlightenment 12; introduction 5–7, 192n.2; nature of light disagreement 14–15; reference frame 8, 21
Next Stage Review (2008) 100
NHS Improvement Plan 2004 99
NHS Management Inquiry Report 1983 99
NHS Reorganisation 1974 98–9
Nixon, Richard 74
Nordhaus, W. D. 133
North Korea 141

Obama, Barak 148, 169, 182
Ohmae, Kenichi 118
Olsen, Johan 27
order: creative complexity zones 56; destructive disorder 56; disorderly social science *26*; Emile (case study) 1–2; foundations of disorderly public policy *28*; foundations of orderly social science *22, 24*; and health 92; and human existence 24–5; Leila (case study) 2–3; Nicolas (case study) 1, 3; order/disorder zones 56; orderly systems 17; paradigm of 12–16, *14, 15,* 20–1; perfection of greater 3–5; pursuit of 5–6; tifling/destructive order zones 56; summary of orderly/disorderly public policy *33–4*

Pakistan 168–9
Parsons, Talcott 22
'Patterns of Global Terrorism' (US annual reports) 173
Perkins, John 135
PESTLE techniques 71
physical complex systems: attractors 38–40; Boolean networks 39–40; explanation 36; limited compressibility/irreversibility 37–8; local interactions/connectivity/simple rules 39–40; local variety/global stability 40; non-linearity 41; strange attractors 39
Plsek, Paul 102

Poincaré, Henri 14
Political Parties (Michels) 27
politics: complexity map of 74–5, *74*; conservatism/socialism 74–5, 81–2; and democracy 76–80; Marxism 20, 81–2; the miserable science 73; the 'third sector' 85–8, 89–91; the 'third way' 80–4;
Politics Against Markets (Esping-Andersen) 120–1
Politics in Hard Times (Gourevitch) 120–1
Pollack, Mark 113
Pope, Alexander 12
post-modernism 25–6, 193n.11
Postmodern Condition: A Report on Knowledge (Lyotard) 25
'Powell Doctrine' 155
Powell, Enoch 98
Prewar Intelligence Assessments about Postwar Iraq (Senate Select Committee) 149–50
Prigogine, Ilya 192n.4
Primary Care Trusts (PCTs) 99
'Principal Challenges in Post-Saddam Iraq' (US document) 149
Prodi, Romano 122
Progress of Nations 1997 (UNICEF) 135
public choice theory 23
public policy 30–1
punctuated equilibrium 134
Putnam, Robert 133–4

Quesnay, Francois 20

rationalism 111–14
Ray, Thomas 42, 194n.11
realism 111
reflectivism 111–14
'Regional Consequences of Regime Change in Iraq' (US document) 149
Ricardo, David 20
'rich picture' techniques 71
Richardson, K. 30
Ricks, T. E. 158
Rihani, S. 50, 132, 133, 135
Rittel, Horst 68
Rostow, W. 132
Rove, Karl 167
Rumsfeld, Donald 155–6, 158
Russia 1, 56

Samuelson, P. A. 133
Sanchez, Ricardo 150–1

Sanitary Commission of Massachusetts Report 1850 95
Sartre, Jean Paul 47
Schelling, Friedrich 25
Schrödinger's Cat experiment 15–16
Scott, James 21, 193n.7
Scowcroft, Brent 149
Seeing Like a State (Scott) 21
sex 43, 194n.12
Shifting the Balance of Power 2001 99
'Silent Surge, The' (Limón) 153
Small States in World Markets (Katzenstein) 120–1
Smith, Adam 20, 40, 135
soft management methods 7, 145, 147
Stacey, Ralph 64
Stalin, Joseph 1
Standards for Better Health (2004) 99
Stanfield, G. 199
State of Denial (Woodward) 155
State of the World's Children 2006 (UNICEF) 135–6
Stiglitz, Joseph 153
Strauss, Leo 156
Strong Democracy (Barber) 76
Sun, the 17
SWOT techniques 71
symbiotic competition 87–9

Taliban, The 166–7, 169, 181, 184
terrorism: 9/11 tragedy 166–8; behaviour of a hegemonic power 170–1; and complexity 178–82, *179*; explaining/reductionist approach 168–9; failure of war on 164–5, 170–2, 197n.3; increase in scale of 173; longstanding threat of 172–3; middle eastern/Islamic 173–8; no single definition of 165–6; not global 169; total international casualties *175*; total terrorist attacks *174*, *176*
Thatcher, Margaret 23, 99
'third way' concept 9
Third Way, The (Giddens) 82
Thompson, Grahame 118
Timberlake, Lloyd 96–7
Toye, J. 132
Transforming Community Services (2009) 100

Uncertainty Principle (Heisenberg) 15
UNICEF (United Nations Children's Fund) 95, 135–6
United Nations Development Programme (UNDP) 130, *131*

Verhulst, P. F. 41

Wahabi tradition 177
Warner, John 156
'Washington consensus' 50, 79–80
'waterfall model' 69–70, *70*, 154
Watford, J. 96
Webber, Melvin 68
Weber, Max 22, 25–6
Westminster model (British) 22–4
WHO (World Health Organisation) 95, 103
'wicked problems' (Rittel/Webber) 68
Wilson, Edward O. 13

Woodward, Bob 155, 159
World Bank 93, 129, 142
World Commission, Environment & Development 144
World Development Report 1993 (World Bank) 93, 142
World Development Report 2004 96
World Transformed, A (Scowcroft) 149

x-y graphs 60–2, 64, 105–7, *106*, 138

Zambia 1–2, 192n.1
Zimbabwe 49

Routledge Paperbacks Direct

Bringing you the cream of our hardback publishing at paperback prices

This exciting new initiative makes the best of our hardback publishing available in paperback format for authors and individual customers.

Routledge Paperbacks Direct is an ever-evolving programme with new titles being added regularly.

To take a look at the titles available, visit our website.

www.routledgepaperbacksdirect.com

Routledge
Taylor & Francis Group